EMPIRES OF EVE

A HISTORY OF THE GREAT WARS
—————— OF EVE ONLINE ——————

VOLUME II

ANDREW GROEN

Author: Andrew Groen
Editor: Cory O'Brien
Layout Design: Annie Leue
Screenshots: Razorien
Portrait artist: Devon Scott-Tunkin
Additional graphic design: Lauren Gallagher (product design,) Bryan Ward
(Strategic Star Map)
Frontispiece: Andrew Ghrist (Instagram: @arghrist_prints)
Propaganda images: Rick Pjanja (98, 101), Poluketes (35)
Special Thanks: Stephanie Spence, Trin Pierse, Matt Lazar, Jim Groen, Taylor Cocke,
Hilmar Veigar Pétursson, Torfi Frans Ólafsson, Kim Swift, Shari Spiro, Carmen
Chavarria, the players and journalists who documented EVE Online in its time, and
the entire EVE Online community.
www.EmpiresofEVE.com
Andrew.Groen@Gmail.com

Ordering Information:
Quantity sales. Special discounts are available on quantity purchases
by corporations, associations, and others. For details, contact the
publisher at the address above.
Orders by U.S. trade bookstores and wholesalers. Please contact:
andrew.groen@gmail.com
Printed in China
Publisher's Cataloging-in-Publication data
Groen, Andrew.
Empires of EVE: A History of the Great Wars of EVE Online / Andrew Groen.
p. cm.
ISBN 978-0-9909724-0-2
1. EVE Online —Gaming—History . 2. Technology—History. 3.
History—Gaming.
First Edition
14 13 12 11 10 / 10 9 8 7 6 5 4 3 2 1

For Ray Bradbury whose address to my
graduating class was a command to write
about what we love, and to love life, and
love it forever.

And for the radiant star who makes loving
life a simple command to follow.

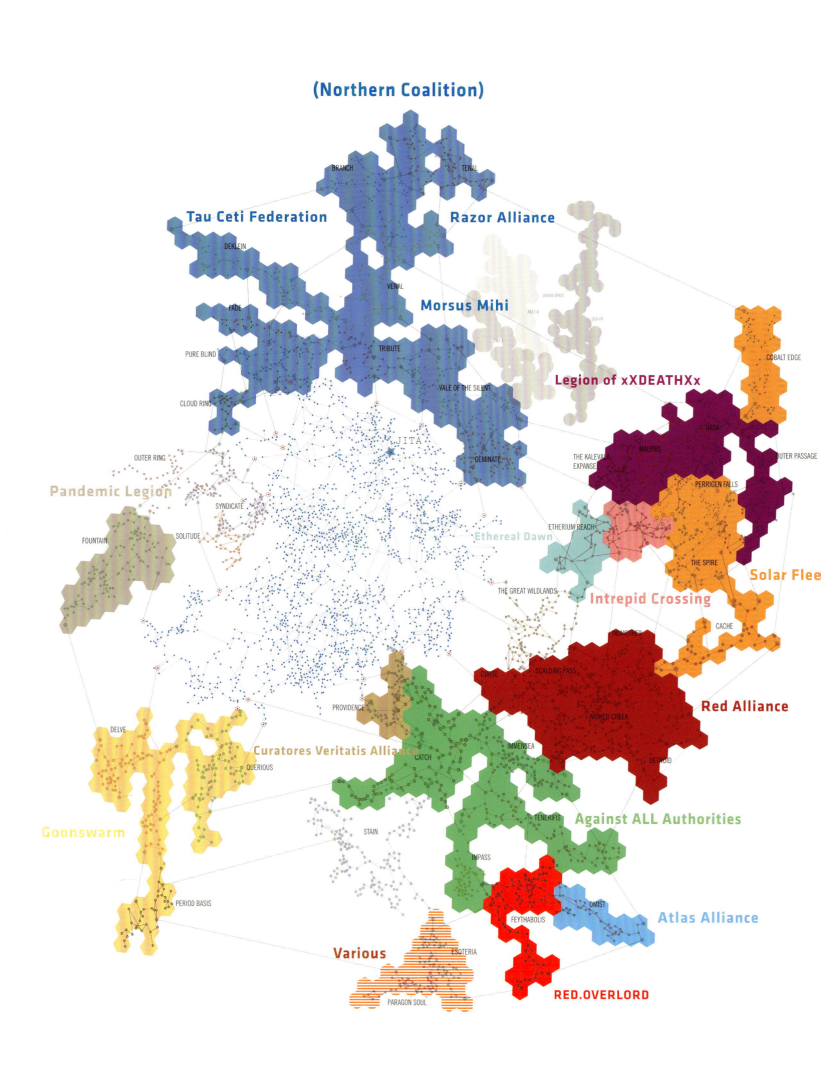

(Northern Coalition)

Tau Ceti Federation

Razor Alliance

BRANCH   TENAL

DEKLEIN

VENAL

Morsus Mihi

FADE   TRIBUTE

JOVIAN SPACE

A821-A   UUA-F4

JYN2-A

PURE BLIND

Legion of xXDEATHXx

VALE OF THE SILENT

CLOUD RING   COBALT EDGE

GÉMINATE

OUTER RING   OASA   OUTER PASSAGE

MALPAIS

JITA

THE KALEVALA
EXPANSE

Pandemic Legion   PERRIGEN FALLS

SYNDICATE   **Ethereal Dawn**

SOLITUDE   ETHERIUM REACH   THE SPIRE

FOUNTAIN   **Solar Flee**

THE GREAT WILDLANDS   **Intrepid Crossing**

CACHE

INSMOTHER

CURSE   SCALDING PASS

Red Alliance

DELVE   WICKED CREEK

PROVIDENCE   **Curatores Veritatis Alliance**

IMMENSEA   DETROID

CATCH

QUERIOUS

TENERIFIS   **Against ALL Authorities**

Goonswarm   STAIN

IMPASS

OMIST   **Atlas Alliance**

PERIOD BASIS   FEYTHABOLIS

Various   ESOTERIA

RED.OVERLORD

PARAGON SOUL

AFTER THE GREAT WAR

Every dot you see here represents a star system in *EVE Online* which is accessible only through the stargate paths that connect them, represented here as lines between each star. Clusters of stars are organized into regions like "Cloud Ring" and "Tribute." The blue stars in the center of the map are called "empire space," but the main focus of this book is on the highlighted outer rim stars known as the "nullsec" regions where player-built empires rule.

In the aftermath of the events of *Volume 1* the political state of *EVE* was drastically reshuffled. Goonswarm abandoned an empire in the southeast to live in the Delve region, the captured home of fallen superpower Band of Brothers. Red Alliance regained its capital in Insmother, but splintered into several factions which journeyed north to claim sovereignty in the Drone Regions led by famed fleet commanders Death and Mactep. With the war over, the Northern Coalition regained all of its lost territory and became arguably the most important power still remaining in New Eden.

TABLE OF CONTENTS

AUTHOR'S NOTE

In the story that follows I have done my best at all times to delineate appropriately between the real and the fictional, but in certain cases I am powerless to stop *EVE* from muddying the distinction. So although the events that follow will at times seem surreal it's worth stating explicitly that *Empires of EVE* is strictly non-fictional.

These are the events of greatest importance as they took place within *EVE Online* from the years 2009 to 2014 as accurately as I could write them as a mere observer of this virtual community.

The story takes place primarily within the digital realm, but EVE is also deeply intertwined with the real world. So the occasional juncture with reality is unavoidable such as when the story leads to that Russian bathhouse in New York City, that bar in Reykjavik, and that embassy in Benghazi.

To keep the line between real and unreal as clear as possible I do not employ storytelling devices like reconstructed dialog or imagined scenes. Anything between quotation marks was actually said or written by the person it is attributed to often in a political speech, forum post, leaked document, or one of approximately 150 interviews conducted during my research.

Similarly, every image of *EVE Online* you'll see in this volume was photographed *inside* the game universe, and minimal photo editing was employed to create these visions of New Eden.

The vast majority of the artwork in this volume was created by a screenshot photographer of rare talent named Razorien who agreed to go on a journey inside *EVE* in 2019 to photograph locations of importance to the narrative. Razorien crisscrossed the stargate network of New Eden in a cloaking ship for weeks to bring us a better view of the wonders of the Tranquility server. The nebulae, planets, and space stations imaged are all really there, and the ships pictured are being flown by actual players.

Amid the shattering of ship hulls and the clacking of hundreds of thousands of keyboards from around the world this book is about the fates of digital societies, a vision of the future of life on the internet set in its past.

Thank you for reading, and I hope you enjoy the story. ●

— Andrew Groen

INTRODUCTION

"What exactly is the population of an online world, and what social forces drive it? One word seems at the center of the issue: Power."
— Raph Koster, Lead Designer, Ultima Online

On May 6, 2003, the tiny Icelandic video game studio CCP Games released a spaceship game called *EVE Online* to stores around Europe. In *EVE*, as it existed in 2003, players piloted starships through a hostile science fiction universe populated by thousands of other players. Though it was predated by an elder lineage of online virtual worlds, this spaceship game had two things that set it apart from its predecessors.

First, it took place in a single online environment. Everyone who played the game inhabited the same, persistent online world as everyone else. That single virtual environment, a cloud of more than 5000 star systems collectively called "New Eden," was online 23.5 hours a day, 7 days a week, pausing only briefly each night for server maintenance.

The other factor that made this game different was a unique sense of object permanence. The starships players could build and fly in *EVE* were costly, and when their hit points reached zero, they were gone forever. That meant every ship, regardless of how big, expensive, storied, or overpowered it was, could always be permanently destroyed. In other words, loss was an integral part of the game design. Every player's kill was another player's death. Every ship and every character had a story all its own, and every story was intertwined.

Because the game of *EVE* is online essentially at all times, events cannot be reset or reversed. Whatever the players manage to achieve—good or bad, purposeful or for lulz—is written into the history of the game, live, as it

happens. *EVE* looks the way it does today because of what happened yesterday, and yesterday can trace its lineage all the way back to the days of the game's launch. Every one of the player characters is linked to every other, even if they play on opposite sides of the star cluster, even if they are from opposite time zones, and even if they played years apart from one another.

In *EVE*, as in the real world, today comes after yesterday and inherits its conditions. If you lost your ship yesterday, you no longer have it today. Nor tomorrow.

As a result of this continuity, the history of *EVE Online* comprises one singularly grand narrative, with millions of characters across—as of this printing—nearly two decades. The majority of those characters scarcely affected the universe at all; most people who try *EVE* quit after only a few minutes. *EVE* is hard. But those who stayed found a universe they could shape according to their personal designs.

That singular narrative has the flavor of the fictional space opera world it takes place within, but because each character is controlled by a real human person the characters are all non-fictional. They are each a real person. Even when a player is actually secretly controlling multiple characters, those characters' choices are still controlled by a human mind, and players' real lives outside the game affect their destinies within it. When we examine the story that exists between all the players we find a spectacularly complex narrative belonging to no individual, but created by the convergence of the entire community.

EVE Online is like that classic parable about a thousand monkeys on a thousand typewriters eventually producing the complete works of Shakespeare. The difference is that the great ape species of *EVE* is the slightly more intelligent *homo sapiens*, and they are far greater in number than a thousand. Perhaps it shouldn't surprise us, then, if the history of *EVE* mirrors *Macbeth*.

THE STORY OF EVE

Some aspects of the game have changed throughout the years. New sectors of space have been added, landmarks have been placed to mark historic events, and CCP has made more than a few graphical improvements. But many things also remain the same. Then as now, the game is set in a cluster of stars called "New Eden"—each star with its own planets and moons, many with space stations, pirate hideouts, or mysterious relics from secretive lost races—for players to explore and encounter each other within. The universe of New Eden is one of extreme beauty and grandeur. Enormous, ribbon-like nebulas streak across each system's skyboxes, and hulking, iron ships slip with perfect grace in and out of the ports of the orbital space stations that serve as player trading hubs.

The star systems in the center of the star cluster are controlled by non-player empires playing out a fictional space opera. Those empires employ a police force called CONCORD which is capable of punishing players who attack other players. In the outer rim of the star cluster, however, far beyond the borders of the NPC empires the star systems have no protections of any kind. Players are free to venture out into "nullsec" (zero security) space and do as they please. They can access rare and valuable materials inside the nullsec asteroid belts and in the wrecks of NPC pirates. They can fight and kill other players. They can institute their own laws if they have the strength to enforce them. They can form groups and become rich, building hubs of commerce.

Previous online games had already shown that players will inevitably seek to form their own virtual societies when given the opportunity to communicate and interact sufficiently. But in the long history of video game development, nobody had ever put so many people in a single confined digital space and forced them to compete for scarce resources. What followed was somewhere between *Ready Player One* and *Romance of the Three Kingdoms*.

THE SOCIAL WEB

Back in 1990, virtual world pioneer Richard Bartle published a seminal essay describing the four chief archetypes of personalities people exhibit in online worlds. The first are the Achievers, who spend most of their time striving to advance further in the game. Next are the Explorers, who search the far corners of the world and test the fine details of its systems. Third are the Socializers, who view the game primarily as a means for interacting with other people and roleplaying other identities. The final type Bartle describes is the one with the most profound implications for the *EVE* universe. Bartle nicknamed this group the Killers, and described their chief governing impulse:

```
"iv) Imposition upon others.

Players use the tools provided by the
game to cause distress to (or, in rare
circumstances, to help) other players."

Richard Bartle, co-creator of MUD1.
From the essay Hearts, Clubs, Diamonds,
Spades: Players Who Suit MUDs, published
in 1990
```

All of Bartle's archetypes found a home in the expansive cosmos of *EVE Online*. But it was the Killers, and their myriad methods of "imposition upon others," who set it on the path to what it is today—for better or worse.

Most of the early players of 2003 *EVE* had already been testers in its incomplete beta state. Those players used that time to establish social networks and figure out advanced fighting techniques and money-earning strategies, so that when the game officially launched those players immediately became important and influential. They knew how to mine and run trade routes in order to get rich quickly on *EVE's* first day, and they knew how to leverage that into wealth beyond the average player's wildest dreams. While they were busily building their fortunes, these players had also unconsciously begun to forge *EVE's* greatest asset: the vast and intricate social web which binds the community together through virtual space and time.

Above: Sunrise over the horizon of the fourth planet of C-J6MT, perhaps *EVE Online*'s most storied star system. Below: The original logo of Ultima Online. Opposite: The cover of the original box art for *EVE Online* in 2003.

But many players were not content to merely make their fortunes as merchants and manufacturers. Instead, they built advanced warships and set up devastating blockades on heavily-trafficked nullsec stargates. These pirates gleefully destroyed and looted the mining ships of brand new players who had no idea yet that player-killing was even possible. Hundreds of these virtually victimized players sent angry messages to CCP Games' game managers. CCP, in turn, calmly explained player-versus-player combat to this first crop of *EVE* newbies, some of whom were playing in their first ever online world.

Some of these players knew that *EVE* was predated by other online games, and that it had inherited their problems. As they always had, the Achievers, the Socializers, the Explorers, and the Killers mutually agreed that their counterparts had no place in the game and were a scourge. They would attack each other and beseech the game developers to change the game mechanics to make it more hospitable to their preferred playstyle and inhospitable to the others. In *EVE*, they do so to this day. And yet, there is a peculiar interdependence between these player groups, despite their mutual animosity, that *EVE* seeks to exploit.

Raph Koster, designer of the legendary virtual worlds *Ultima Online* and *Star Wars Galaxies*, put it well in his observations on Bartle's findings:

"The fascinating part of the essay is where Bartle discusses the interactions between these groups. Killers are like wolves, in his model. And therefore they eat sheep, not other wolves. And the sheep are the Socializers, with some occasional Achievers for spice. Why? Because killers are about the exercise of power, and you do not get the satisfaction of exercising power unless the victim complains vocally about it. Which Socializers will tend to do.

Further, Bartle pointed out that eliminating the Killers from the mix of the population results in a stagnant society. The Socializers become cliquish, and without adversity to bring

communities together, they fragment and
eventually go away. Similarly, Achievers, who are always looking for the
biggest and baddest monster to kill,
will find a world without Killers to
be lacking in risk and danger, and will
grow bored and move on.

Yet at the same time, too many Killers will quite successfully chase away
everyone else. And after feeding on themselves for a little while, they will
move on too. Leaving an empty world."

Raph Koster, Lead Designer of Ultima
Online, one of the primary design in-
spirations for *EVE Online*
May 7, 1998

The magic of *EVE* is that it seeks to find a balance
between these types of players, and provides a grand interstellar stage for that human social ecology to play out.
CCP Games vowed to judge no one's playstyle. The universe was a sandbox, and whatever you decided to build
with the sand was your choice. It was even your choice
whether to build with the sand or throw it in someone's
eyes and smash what they were building. Where in previous online games Killers were reviled as literally immoral,
EVE Online offered them a place within its ecosystem and
allowed them to fulfill their role as catalysts for change.
It asked the community of the game to govern itself as
much as possible, only stepping in themselves if there was
outright cheating or a threat of real world violence. At
the same time, the developers nudged *EVE* toward war
by creating systems for collaborating with friends and
uniting against enemies.

In 1998, in response to a dispute between Killers and
Socializers within his own community, Ultima Online's
Raph Koster wondered what he was truly observing:

"What exactly is the population of an online world, and what social forces drive
it? One word seems at the center of the
issue: Power. The conflicts that arise
are there precisely because competing
agendas (and often, as in the case of
the playerkillers versus the roleplayers, competing play styles) attempt to
exercise power over one another. [...]

We must have playerkillers in UO,
because the world would suffer if we
did not have them. But they also must
be channeled, so that their effect is
beneficial, and not detrimental. [...]
It's largely about perspectives. The
issue for the Killers is whether they
will gain the wider perspective and
cease to be "virtually sociopathic."
And the issue for the Socializers is
whether they will recognize that the
Killers are a part of their society too,
and not always a bad one.

The thorny issues that then remain
are the nitty-gritty of virtual community building: how do we govern in a
world of anonymity? How do we police,
and who polices, the players or the game
administrators? What sort of punishment
is appropriate for virtual crime? What
sort of punishment is even possible for
virtual crime?"

Raph Koster, Lead Designer, Ultima
Online
May 7, 1998

Reading Koster's and Bartle's writings, it is hard not
to believe that they somehow knew the destiny of *EVE*
before it was even built. It seems that the patterns at play
in *EVE Online* are hardcoded into our systems, similar to
the patterns that govern the motions of cities. However,
just like cities, the design, layout, and governance of a
virtual world play an essential role in shaping the people
and their quality of life. Each person is undeniably an
individual, and yet—knowingly or not—they fit inescapably
into a grander human pattern.

The ability of *EVE Online* players to exert power
over one another incentivized piracy, plotting, and bold
schemes. The risk and danger caused by this forced people
together and lead them to form groups for safety. The
groups they formed almost always coalesced around certain
types of gameplay: socializing, exploration, achievement,
and—most infamously in this universe—"imposition upon
others."

EVE is an experiment that has been evolving since
Day 1, and we continually observe new behaviors. In
many ways *EVE Online* has always been an astoundingly

Above: A player is killed in a confrontation leaving their ship, capsule, and the clone inside permanently destroyed.

forward-thinking game. It pioneered a number of technologies and design practices that are still being jealously studied by CCP's competition to this day. In 2003, for instance, the server hardware that operated *EVE Online* was so advanced that CCP Games needed to lobby to get an exemption from the United States Military in order to export it from the US to Iceland. Almost twenty years later, its technology is still capable of amazing feats, yet its design is full of old truths the gaming industry has forgotten.

The design of the game meant that the community was constantly tearing itself apart and putting itself back together again.

PRE-HISTORY

The story of *EVE* is not something that can be written down in full. There are more characters in that story than there are words in both volumes of this book. The community's travails, told in full, could fill a thousand books with a thousand pages, each story intimately interwoven with the others. It would likely be impossible to read because all of the stories happen concurrently.

Empires of EVE: Volume 1 summarizes the story of *EVE's* first six years of existence, focusing on the player-conquerable "nullsec" regions of space. The *EVE* community was generously patient with my storytelling as I tried to condense a 4-dimensional story onto the 2-dimensional page, necessarily leaving out volumes of detail and nuance. I'll have to ask for more of that patience as I attempt to summarize that summary:

On May 6, 2003, when *EVE* first launched, the vast majority of players were still learning how to play the game in high-security space, and the few who were brave enough to venture outward found hundreds of star systems that were utterly vacant. A steady stream of players began to make their way into the nullsec regions, exploring their outer reaches. Nullsec was famously spare and lonely in those early days, with the exception of the occasional merchant ship taking the risky road in search of profitable minerals, and the pirates who preyed on them. On occasion, those pirates found choke points in the stargate map that could be used to shut down entire regions of commerce. The wolves essentially found a mountain pass that funneled the sheep directly to them. One such incident by a pirate group called M0o very nearly broke the entire game. Early nullsec was ruled by these roaming pirate gangs.

In the far more populous interior star systems called empire space, many miners and traders (nee: achievers and explorers) became early billionaires who could afford to construct the game's most expensive ships. It was only a matter of time before the industrial billionaires realized they could use their superior weaponry and pay mercenary pilots to force out the talented, but underfunded, pirate fighters. The cleverer billionaires simply bribed the pirates directly to become their "military." These industrialists offered the player-killers a new story: instead of being vagabonds who robbed travellers, they could be soldiers for a cause. In exchange for coexisting with the peaceful players inside of a structure of rules,

THE GREAT SCAM

The most famous story to come out of *EVE*'s early years is a tale known as "The Great Scam." It's a 13-part opus that is partially real and partially embellished (the author eventually admitted privately that the final page of the story was fiction they had added to give the story a tidier ending for their college creative writing class.) Nonetheless, the story remains famous within the *EVE* community because it illustrates something fundamental about how *EVE* works and where its unique magic really comes from.

"The Great Scam" recounts in detail how a player named Nightfreeze raised their first million ISK (EVE's in-game currency—a million was a small fortune in 2003, but is a pithy sum to a modern player) by taking out a loan from a more successful friend and buying an industrial ship so they could start a trading business. Freshly outfitted with their new ship, Nightfreeze had an unfortunate run-in with a Killer from low security space, "DanielSan," who cornered Nightfreeze's ship, brought low the ship's shields and armor, and then demanded payment to let Nightfreeze escape complete destruction. But Nightfreeze had invested everything into this ship, and had no cash left to pay the ransom. As Nightfreeze pleaded for mercy the pirate launched a final missile and Nightfreeze's new industrial ship was destroyed. Nightfreeze was at first inspired to fight back and rid the game of player-killers and bullies, but was eventually beaten and forced to quit the trading business. All the while, the feud between Nightfreeze and DanielSan escalated, and eventually, Nightfreeze was tempted to adopt some seemingly immoral practices.

With the business venture ruined, Nightfreeze lost all sympathy for the rest of the players of *EVE*, deciding that scruples were a disadvantage in a game where unscrupulous Killers ran free. Nightfreeze devised an elaborate pyramid scam which would lure in new players and leave them penniless.

This much of the story is true. Nightfreeze later said the encounter with the pirate DanielSan late one night in 2003 actually did happen, and the fraudulent investment scheme really was planned. But after that, the true story ends.

The fictional Nightfreeze in the creative non-fiction assignment executed the investment scam to perfection and became fabulously wealthy, a morally-compromised kingpin in a starry dystopia. The real Nightfreeze stopped playing the game before they got a chance to execute it because their best *EVE* friend moved on to a different online game and Nightfreeze stopped caring.

It's a quaint story by today's *EVE* standards. Players with stories like this in the modern *EVE* community are listened to with a patient sympathy, because most everyone has lived some version of that descent.

But this story is fascinating nonetheless, because it raises an important question: if *EVE*'s game design can provoke such intense animosity between a handful of pilots over the few weeks that NightFreeze's story encompasses, what has it done to its millions of pilots over its 16+ year lifespan? This book is the story of some of those years, and of players who didn't quit before they executed their master plans.

the Killers got better weaponry and the opportunity to be part of larger groups and bigger operations. However, the combination resulted in constant ideological friction. The two groups could never agree on rules governing behavior, because they weren't just people who wanted to play different ways; they were people who saw the game's purpose differently.

This is how the first makeshift governments of *EVE* were formed. Roleplayers, traders, and industrialists (Socializers, Explorers, and Achievers) allied with pirates (Killers) out of mutual advantage. When friction inevitably emerged, they formed council governments to resolve those disputes. These alliances were a cutting edge social

technology, and the best of these groups gained immense wealth. As these alliances grew rich, they expanded. They used their new wealth to offer incentives to new recruits by spamming local chat channels and offering free ships and regular paychecks. But the most important thing they offered their pilots was the chance to be a part of something larger than themselves: a community. These communities or "corporations" attracted like-minded players, and encouraged other established communities to form alliances. These corporations' quickly expanding populations spread out over hundreds of largely vacant star systems, which they controlled by using starship patrols to hunt down pirates.

One of the most famous inter-faction conflicts in these early days was a war between a roleplayer named Jade Constantine and her corporation Jericho Fraction on one side, and an ultra-wealthy industrialist named Ragnar Danneskjold and his corporation Taggart Transdimensional on the other. Jade Constantine was a master socializer—one of the greatest in online gaming history—who used her voice to paper the forums with screeds about how Ragnar had been caught associating with rogue pirate factions. Their enmity eventually exploded into a series of civil wars which broke their organizations and created new, more complex ones.

From the ashes of their war was born the "Phoenix Alliance," a broad coalition of corporations which, though each had its own individual culture, were united under a strong leader named Halseth Durn. Guided by their leader, this organization amassed incredible wealth and power in the northern reaches of New Eden. Durn, an accomplished *EVE* diplomat, achieved this great political feat by bringing together all three major alliances of the north into a single coalition under the "Northern Alliances Security Treaty."

But that early coalition was undone by a scandal among the leadership (which seems to be unavoidable in groups of this scale) and collapsed during the "Great Northern War." Both sides of the conflict lobbied the other player groups of the star cluster to side with them and enter the war on their side. Over months of battles the conflict grew to encompass nearly a third of the zero security areas of *EVE Online*. Player alliances that were too far away to send their members to fight would send ISK, ammunition, and ships to their allies.

There had always been small-scale conflicts in *EVE*, but the Great Northern War in 2004 was a wake-up call. It forced the young *EVE* community to reckon with three truths: 1) it was possible not only for a virtual society to form, but for a geopolitical framework to be established, 2) that very geopolitical framework could potentially ensnare every major group in the game into a single all-encompassing political net, and it foreshadowed 3) that this net could drag all the alliances of *EVE* into a massive game-wide battle.

The end of The Great Northern War dismantled the power structure of the north and heralded a new dynamic for *EVE Online* partially stimulated by its ever-growing playerbase. No longer was every alliance its own discrete group of corporations. Each of them was intimately intertwined with the others in what were now being called "coalitions." In essence, everyone needed to be a part of a larger coalition of powers in order to defend themselves against the threat of their enemy's coalition of powers. As one group adopts a new social technology, so must the others or perish.

The players formed into "corporations." The corporations merged into "alliances." The alliances formed into "coalitions." The scale of diplomacy and conflict soared to ever more impressive organizational heights. The wars that occured could ensnare tens of thousands of players, and yet the dramas that drove these organizations to war were almost always small, human-level controversies between just a few people who happened to command vast influence.

The most important event to occur in this era was the founding of an organization known as Band of Brothers (BoB). In the tempest of the Great Northern War, the social networks of the combatants collided, and several of the military organizations in that conflict realized they would be more powerful if they could come together to form a singular, unstoppable military alliance.

In his 1990 paper, Bartle described what happens when Killers encounter one another, and he might as well have been describing the formation of Band of Brothers. "Killers try not to cross the paths of other killers, except in pre-organised challenge matches," he wrote. "Part of the psychology of Killers seems to be that they wish to be viewed as somehow superior to other players; being killed by a Killer in open play would undermine their reputation, and therefore they avoid risking it."

At first, this group was merely a roving military group, but its leader—a Swedish heating and cooling repairman who went by the name SirMolle—soon set his sights on territory of his own, and Band of Brothers went on a mission of conquest.

Their first wars were with well-selected, soft-targets meant to allow Band of Brothers a chance to practice large-scale warfare. Dreadnought-class siege ships had just been patched into the game, and their heavy cannons became the go-to tool for grinding down crucial enemy infrastructure. BoB used soft targets to test out these new and expensive ships and observed how they operated in the field. As their victories continued, the alliance grew ever more bold.

The fact that Band of Brothers played the game at the very top level attracted many players who desired to join its ranks, including some of the game's developers. Some of them saw this as an invaluable opportunity to experience their own game live, and see how it actually worked. Fueled by this new membership, and the spoils of his victories, SirMolle soon came to see himself as a great conqueror.

In one of the more surreal moments in *EVE Online* history, SirMolle was interviewed by the New York Times:

"Our goal is to control all of *EVE*," he told New York Times reporter Seth Schiesel. "It's totally impossible to claim all of *EVE* physically. But it's possible to control the people. It's possible to control the alliances, be it by economic means or fighting means or political means. That was the goal and that is the goal."

He and his inner circle devised Operation Clockwise, a cocksure plan to dominate all of *EVE Online* by moving around the star cluster, destroying or subjugating their enemies one by one. He even joked about eventually acquiring enough power to kill every NPC, conquer the game, and then set off to Iceland to conquer CCP Games' headquarters near the docks in foggy Reykjavik.

In the early days, Band of Brothers' most formidable opponent was a large group of players known as Ascendant Frontier. This group was comprised mainly of former players of the older online space game Earth & Beyond. When Earth & Beyond was shut down by its publisher Electronic Arts, many of that game's largest organizations joined *EVE Online* as a group. Earth & Beyond was a game far more tuned for Achievers than Killers, and thus the group that joined *EVE Online* was a group focused on building as much as possible as quickly as possible. Band of Brothers controlled the west, which it built into a fortress. Ascendant Frontier controlled the south, which it built into a prosperous economic hub. The north had recently banded together in a defensive pact called the Northern Coalition, to protect themselves against the clear-and-present threat of Band of Brothers.

Ascendant Frontier's leader was CYVOK, a satellite communications expert for the US military who ran Ascendant Frontier in his off hours. CYVOK was a calm, well-liked alliance leader who was focused on building a community of players, and a thriving hub of commerce in the nullsec territories.

In the summer of 2006, CCP Games introduced the all-new Titan-class ship, a colossal interstellar juggernaut armed with a devastating "Doomsday" weapon that could wipe a battlefield clear of enemies in a single blast. CYVOK made the decision that Ascendant Frontier would be the first to build one. It took three months for CYVOK and his closest allies to build the ship. Much of this time was consumed by counter-espionage, as the other nullsec alliances would have banded together to prevent him from acquiring the mightiest ship in the game if only they had known where he was building it.

On September 25, 2006, CYVOK's Titan was born. The reveal shocked the *EVE* community and CCP Games alike, because it had been built months before CCP Games believed it should have been possible to do so. CYVOK said it was "three months of the most boring thing I have ever done in *EVE Online*." He named it "Steve."

STEVE

Unknown to CYVOK, SirMolle was in the midst of his own Titan project—"Darwin's Contraption"—and was furious that his 13 kilometer flagship would not be the first in the history of the universe. SirMolle

Opposite: The Caldari Navy Assembly Plant in Jita is generally the most heavily trafficked location in *EVE Online*, thanks largely to its convenient location. Above: Viewed from an exact distance, a ringed planet and its moon appear the same size. Below from left to right: Early *EVE Online* leaders Halseth Durn, Jade Constantine, and SirMolle.

reasoned that if he could not be the first to build a Titan-class ship, then he would be the first to destroy one. SirMolle announced Band of Brothers' intent to invade Ascendant Frontier just two days after Steve was born.

Seventy-seven days later on December 11, 2006, a battle broke out near the front line of the Band of Brothers assault. Two relatively small fleets engaged each other and skirmished for control of a system in the devastated no-man's-land between the two virtual empires. And lo, Steve did enter the battle to confront Band of Brothers. The BoB fleet scattered at the sight of the behemoth, but not before two BoB ships were obliterated by the Titan's primary "Doomsday" weapon.

CYVOK warped Steve to a safe spot in deep space, and—the legend goes—he switched to a different PC in his home to try to clear up some system latency issues, forgetting that his ship would remain in space after he had destroyed those two BoB ships. After logging off, a BoB scout conducting a random sweep found Steve the Titan floating lifelessly through space without an online pilot.

Upon hearing the scout's report, SirMolle ordered his entire dreadnought fleet to warp to the Titan in deep interstellar space, and begin tearing into its hull.

CYVOK logged back in and discovered Band of Brothers' dreadnought fleet staring back at him, and logged back out of the game. Just days later he quit *EVE* forever, and without his stewardship, Ascendant Frontier—a

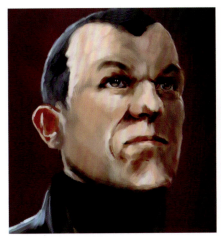

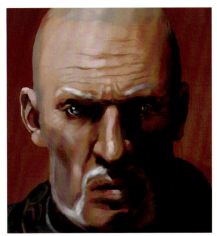

community of thousands of gamers—fell apart completely and was steamrolled by the experienced, well-coordinated BoB fleets.

For most of *EVE Online*'s history it was believed that Steve had been destroyed because CYVOK logged off too soon after the battle—that his ship never completely logged out of the game due to the recent combat, and it was discovered by a fluke scan by a BoB scout. CYVOK has always maintained that this story is false. He says the murder of Steve was an inside job, perpetrated by someone with developer access who abused their developer powers to find CYVOK's ship when it should have been impossible to (since he had teleported it to a deep space location that should have taken any BoB ship 23-hours to fly to unless they had the secret jump coordinates.) However, extraordinary claims require extraordinary evidence, and in this case CYVOK acknowledges that evidence is nearly impossible to come by because he wasn't recording his gameplay at the moment it happened. Most of his *EVE* contemporaries don't believe him, but I spoke privately on one occasion with a former CCP developer who does.

CYVOK quit the game a few days later, but still says he loves *EVE* and thinks it's the greatest MMO ever made. He does not allege that it was a conspiracy. He thinks it was just another classic case of a random Killer—who happened to have developer privileges—exercising power over an Achiever he thought was ruining the game by flooding the economy with cheap ships, and dominating nullsec with a Titan he shouldn't have been able to build.

Regardless of how it actually happened, the result of Steve's destruction was the same. CYVOK left the game, and Band of Brothers swept through Ascendant Frontier's territories, claiming everything for themselves. With the fall of Ascendant Frontier, fully half of *EVE Online* was now conquered by a single group of a thousand pilots and the many vassal alliances it installed.

But the same fear that kept BoB's vassals loyal also inspired resistance from the other powers in nullsec. Band of Brothers' expansionist goals were obvious to anyone who was paying attention, and the other alliances knew that they were bound to share Ascendant Frontier's fate if they didn't stick together. On top of that, whispers were circulating that members of Band of Brothers were actually cheating devs.

Above: The leader of Ascendant Frontier in 2005, CYVOK. Below: The logo of Ascendant Frontier. Opposite: The cracked silhouette of the wreck of Steve the Titan.

And so the disparate leaders of nullsec put their differences aside and formed a mutual defence pact against Band of Brothers. The battle lines were drawn. You were either with Band of Brothers or against it. BoB vs *EVE*.

As SirMolle prepared to invade and topple the Northern Coalition, scandal struck. A hacker working as a rogue intelligence agent broke into Band of Brothers' out-of-game forums and communications channels and discovered evidence that a member of Band of Brothers actually was an *EVE Online* developer who he could now prove had illegitimately used his administrative powers to help Band of Brothers acquire a cache of highly valuable "Blueprint Originals" which created a persistent flow of income for the alliance.

When the hacker revealed this to the *EVE* public, there was a community revolt. This, many people thought, was the perfect explanation for Band of Brothers' massive success. To them this developer wasn't a lone wolf, it was merely the first wolf they'd managed to catch. BoB's political enemies jeered, calling them "Band of Developers," while urging justice-minded *EVE* players to join their crusade to undo the corrupt BoB regime. Just because history can't be reversed, they thought, doesn't mean it can't be unmade. In a staggering display of cooperation, dozens of alliances came together with the goal of overthrowing Band of Brothers and destroying its means of organization.

The coalition of alliances opposing Band of Brothers struck quickly over a massive area. Invaders flooded into Band of Brothers territory, but BoB proved remarkably resilient to the onslaught. One by one the attackers were driven back as Band of Brothers miraculously stabilized against a horde of foes.

SirMolle recruited the soldiers-for-hire Mercenary Coalition to strike back in the north, and with staggering effectiveness the whole of the north was captured, ceded to the control of Band of Brothers.

TITAN // AMARR EMPIRE
AVATAR

"For those who did not play *EVE* at the time, it is difficult to comprehend just how powerful BoB had become," wrote James 315, one of the few players who documented the events and characters of this early era. "Its core was made up of the oldest and most successful PvP alliances in history, some from the beta. It had never lost a war. Ever."

By now, 2007, the war effort to unseat BoB had backfired spectacularly. Rather than containing Band of Brothers, the attack had unleashed its true potential. Though Band of Brothers had been implicated in scandal, it was still nullsec's preeminent military organization and included some of the oldest and most successful combat-focused groups. Ironically, given its earlier crusade against Steve, BoB soon became widely despised for building more than a half dozen Titans of its own. These, coupled with the expertise of the empire's pilots, made them functionally impossible to fight. Two-thirds of the game now rested under the control of BoB and its allies, and "Operation Clockwise" ticked on toward midnight.

In the East stood the last true hope of stopping Sir-Molle from conquering every star system of nullsec: RedSwarm Federation, a union between the Americans of Goonswarm, the Russians of Red Alliance, and the French of Tau Ceti Federation, which came together with the singular goal of dismantling the BoB power bloc.

The Goons were an anomaly in *EVE Online*, populated as it was mostly by pompous space dictator personas and grand war fleets. Its pilots fit into the "killers" category, no doubt, but they were more famous for griefing and luring other players into scams (activities they considered perfectly acceptable gameplay) than for stellar conquest. And since they recruited their membership entirely from the SomethingAwful.com community, the leaders of New Eden often saw them as outsiders to be feared.

The same was largely true of the Russian community, whose members were vilified as cheaters involved in Real Money Trading operations (the selling of in-game items/ISK to other players or brokers for real world currency.)

The geopolitical conflict was fueled further by a deep multilateral grudge. Two years earlier, SirMolle had bullied Goonswarm, and sought to drive the Goons out of *EVE* after one of its members made a stick figure comic mocking the real world death of a BoB pilot who had died in a motorcycle accident (and whose father was also an active member of BoB.) SirMolle famously declared on the forums, "There are no goons," and camped the Goon headquarters in the region of Syndicate.

The Russians of Red Alliance had faced similar treatment by many of BoB's allies, who had captured and occupied Insmother, the Russian "ancestral homeland" in *EVE*. The effort by BoB and its allies to drive Red Alliance and Goonswarm out of the game instead gave them common cause. The glue that bonded them was the mutual experience of being told "your kind isn't welcome here." Even when it happens virtually, this is not a feeling human beings take kindly to.

RedSwarm Federation was a "coalition of pariahs"—as one leader would later call it—formed specifically because of their mutual outsider status, having been pushed to the brink by *EVE's* powers-that-be.

The tremendously bitter war between RedSwarm Federation and the Greater Band of Brothers Community raged for months. At times, each side seemed to have

victory within its grasp. But at last, a flurry of inspiration combined with a sudden shift in game mechanics to change the course of the war. As SirMolle tried to break through RedSwarm's faltering defenses, he forgot that CCP had recently released a patch that nerfed Titans. He was trapped, and Darwin's Contraption—his prized Avatar Titan—was destroyed. It proved a critical turning point. Over the next 18 months, BoB's gains were slowly undone, and as each of SirMolle's next three Titans went nova in succession, the illusion of BoB's mastery unraveled.

THE DELVE INVASION

In time, RedSwarm shocked the *EVE* community by using its enormous membership to overwhelm Band of Brothers and force a full retreat into BoB's home region of Delve. Inside his "Fortress Delve," SirMolle made his last stand. Siege defenses were anchored in every possible location in every star system.

With Band of Brothers and Mercenary Coalition forced to hole up in Delve, the Northerners rose again, reconquered their home, and in a few weeks were able to join in the final assault. Together, the Northern Coalition and RedSwarm Federation launched a full-scale attack on Delve and the battle of the decade was begun. Tens of thousands of ships were destroyed in the calamity—faster than the industrialists of *EVE* could build them—leading to resource scarcity all across New Eden.

But as the lag cleared, Band of Brothers was still standing. Its enemies complained that the game was imbalanced as they retreated from the battlements of SirMolle's Fortress Delve. One by one, the alliances that comprised the Northern Coalition and RedSwarm Federation gave up hope of capturing Delve and returned home.

Spared ultimate destruction, Band of Brothers rebuilt and regained much of its former strength, but the conflict was far from over. A small squabble between two unrelated groups soon dragged in several other alliances, which eventually dragged in both BoB and Goons, and began the inevitable next phase of what was now becoming known as "The Great War."

Both sides braced for yet another long, violent siege of the Delve region. But it was treachery, not warfare, which decided Band of Brothers' fate. A high-level player inside Band of Brothers named Haargoth Agamar—a director with access to the corporate accounts and ship hangars—had secretly been growing to hate his own alliance. Agamar defected to his former rivals in RedSwarm Federation, declaring that "BoB is all emo now and can't even stick to a fucking plan."

But Agamar did not simply defect. His treachery consisted of nothing less than the complete destruction of his former alliance. With his final act as a Band of Brothers director, he used a little-known loophole to shut down the shell corporation that served as the umbrella organization for all of the alliances in Band of Brothers. It was as though there was a massive power switch that turned off Band of Brothers as a whole. Because of the way the gameplay mechanics in *EVE* work, this meant that every single one of Band of Brothers defensive emplacements were shut off the next morning when the server tried to determine who owned them and found no alliance on record named "Band of Brothers." Goonswarm quickly re-registered an alliance called "Band of Brothers" so that the original BoB could never regain its own name.

The old BoB corporations moved quickly to re-register themselves as a new alliance—their new name was Kenzoku, a Japanese word meaning "a family of people who have made the same commitment"—but in this era of *EVE* it took 4 weeks for defenses to reach their full power and potential. This meant that for the next month SirMolle's impregnable citadel was completely exposed.

It had been 18 months since the beginning of hostilities between Band of Brothers and the coalition that RedSwarm assembled to destroy them, and in this moment RedSwarm Federation saw an opportunity to end it forever. The leader of RedSwarm—Goonswarm CEO Darius JOHNSON—decided that this was to be the final battle. He ordered that every member of his alliance should pack up everything they owned and prepare to move directly into Band of Brothers' home region. Every piece of infrastructure in the former Goonswarm territory was to be torn down or destroyed.

To say that the fighting that ensued inside Band of Brothers' home region was fierce would be an extreme understatement. For weeks on end there were thousands of people fighting inside the 80+ star systems that comprised the compromised Fortress Delve. But the result was inevitable: After two weeks of the most fierce and lengthy battle in the history of online video gaming Delve was taken, and Band of Brothers was destroyed.

Opposite: The leader of RedSwarm Federation, Remedial. Above: RedSwarm spymaster, The Mittani.

It was mid-2009, and Band of Brothers had dominated the game since 2004. Nobody knew what a world without BoB even looked like. Some organizations had no other goal or purpose for being except fighting Band of Brothers. They were left rudderless, confused even, in the wake of the relatively sudden death of their great enemy.

The defeat of Band of Brothers ushered in an entirely new era in *EVE Online*. RedSwarm Federation and the Northern Coalition were left as the only two credible power blocs in nullsec, and the entire fate of the game was left up in the air. Nobody knew what came next, but anyone could see that a new age had dawned in *EVE Online*.

COLD, DARK, AND HARSH

Early in *EVE's* life, a community manager summed up the game's ethos with one of the most enduring descriptions that defines the game to this day.

"EVE isn't just designed to look like a cold, dark, and harsh world," they wrote. "It's designed to be a cold, dark, and harsh world."

It's an understatement. *EVE* is a social bloodsport in which the cost of failure is often the destruction of

Above: A monument erected in C-J6MT by CCP Games to honor the history of this much-contested star system which was so often at the center of *EVE's* history.

your community. And because there are no limits on how much the player can apply themselves toward their goal, gameplay at the peak of *EVE* sometimes involves drastic extremes of human behavior like rage, deceit, propaganda, character assassination, and weaponised racism and misogyny.

In creating space for the best in us, the *EVE* ethos also created space for the worst in us, and there are many scars and much graffiti on the history of *EVE* to prove it.

However, there are many people here who are genuinely warm, bright, and kind. For them, the fight against the cold, dark, and harsh is the entire point of their time in *EVE*. The leaders of these organizations of players describe their experience with striking similarity. Many of the corporation leaders I spoke to describe a feeling of deep exhaustion. It is often only through sheer zeal and community loyalty that many of the organizations described in this book manage to survive against the

crushing natural entropy of the *EVE* universe. If there's one thing that unites most *EVE* players I've met it's a shared love of darkness. Either fighting it or revelling in it.

Every one of those leaders has told me they would quit *EVE* if they could; some of them desperately want to. They're tired, but they've been here long enough to know how rare their light is in this place. Many of them have been here since 2003, and they've endured 17+ years of cruel conquerors and cynical con artists becoming increasingly sophisticated and diminishingly discerning.

With their newbies chattering on comms in the glittering darkness, the humble corporation CEO waves a torch to keep *EVE's* nature at bay a while longer, wondering how long they have until the fire gets too dim to scare back the darkness.

Which it eventually will. In *EVE*, nothing is more certain.

THE COUNCIL OF STELLAR MANAGEMENT

In response to the outcry over the revelation that an employee of CCP Games had been caught cheating, CCP Games took a number of actions to secure the integrity of the game and prevent further disconnects between the company and the community. One of the changes that was made was the introduction of a player-led council to serve as advisors to CCP called the Council of Stellar Management.

Every six months the player community would vote for representatives, and the winners would be flown from around the world to meet at CCP headquarters in Reykjavik to discuss CCP's agenda for the game and debate its future. CCP Games was growing more and more aware that *EVE Online* had progressed from a video game into something more closely resembling a society. The company had always believed that the players had two primary rights in the game world. 1) The right to be free of undue external influences in the virtual society (the right to not be threatened or assaulted in the real world.) 2) The right to unlimited interaction with other individuals. In other words players were free to do anything and everything within the bounds of the Terms of Service. Things like fraud, deception, or destruction of property were in-universe problems to be solved by the virtual society.

Now they added a third right. 3) The right to have influence on how society is legislated.

"The goal of CCP is to provide *EVE's* individuals with societal governance rights," reads a sort of manifesto published by CCP. "Once elected, the responsibility of these representatives will be to uphold the society's views as best they can via direct contact and dialogue with CCP."

CCP's admirable idealism would be put to the test over the coming years as the "right to have influence on how society is legislated" and the "right of unlimited interaction with other individuals" collided in one of the world's largest ever experiments in digital democracy.

The first Council of Stellar Management was elected on May 21, 2008 with 14 players awarded seats on the council including Jade Constantine and Darius JOHNSON.

TRANQUILITY

Never was this principle of inevitable entropy more clear than in the case of Band of Brothers, who had once looked as if it had broken the game and would be impossible to stop from establishing a system of complete control over all of New Eden. This was the story of *Empires of EVE: Volume I.*

Empires of EVE: Volume II is the story of the next stage in that unending cycle: of those who came to power amid the downfall of BoB, and how those powers grappled with an entirely new dynamic in nullsec. The balance of power had changed. The game mechanics of nullsec were set to change. The Internet itself was changing outside and around the game, and *EVE Online* was on the brink of a golden age.

With the conquest of Delve complete and the final BoB base at 49-U6U captured, SirMolle and his lieutenants were chased back to empire space in defeat. Goonswarm CEO Darius JOHNSON resigned from his post and with his final proclamation trolled triumphantly, "Delve is Goons. Anime is cartoons. Welcome to the most profitable and heavily fortified region of space in the galaxy. Delve is impossible to invade. We know because we've tried. Our bitter struggle is finally over, and at long last our revolution is complete."

Much of the story that follows deals with the seismic effects of this revolution, and the farflung destinies of the groups who managed to achieve it.

It's also the story of how the victors managed to blow it completely. ●

REDSWARM FEDERATION

"Unity succeeds division, and division follows unity.
One is bound to be replaced by the other after a long
span of time. That is the way of things in the world."
— Luo Guanzhong, Romance of the Three Kingdoms

Sometime in mid-June 2009, as the Band of Brothers defense was in its failing stages, there was a demonstration of sorts in *EVE Online*.

A coalition of player pilots from RedSwarm Federation and the Northern Coalition gathered its combined fleet of Titan starships—27 in total—for something like a military parade. The star system was 49-U6U, the site of BoB's final outpost.

The 27 ships—more than a quarter of all Titans ever built in *EVE Online* at this point—organized themselves into an arch around a single dummy ship just outside the outpost. A sight like this would have been unthinkable at the beginning of the Great War two years ago, when there were only four Titans anywhere in *EVE* and each was a tightly controlled strategic asset. But the Great War had created the infrastructure to push military production to previously unthinkable heights, and now 27 of them were drifting in space, waiting to destroy an empty, stripped down capital ship hull. Just for fun.

"The camp was a show of force, a show of strength if you will," a fleet commander named Viper Shizzle of the alliance Pandemic Legion told EVEOnline.com. "Sitting 27 Titans in a hostile system 100km off of a station has no strategic value other than simply [saying] 'we can do this and you can't stop it.'"

A countdown began, and as it reached zero the pilots fired their Doomsday weapons by type. First, the Avatar-class Titans drowned out the light of the nearby star with nova flares of holy Amarrian light. Next came the Erebus ships, which bathed the entire area in a sea of slow

red flame. Third, the Leviathans summoned pulsing blue wormholes. Finally, there were the Ragnaroks. The last four Titans erupted with a scatter of hundreds of missiles firing chaotically in every direction, which detonated in a lumpy ball of digital smoke and flame some 1000 kilometers across (or the width of your thumb, depending on how closely your camera was zoomed.)

The Titan fireworks display was a statement to BoB—or 'KenZoku' now that Goonswarm had stolen its name—and to the rest of the *EVE Online* community meant to show beyond any doubt that BoB's time had come to an end.

THE JESTER

If the Great War were told on an ancient papyrus scroll it would go something like this: After telling a despicable joke, the Court Jester (Goonswarm) was beaten and jailed by The King's guards (Band of Brothers.) Stewing in prison, the vengeful Jester concluded that the funniest revenge possible would be to break out of jail, unite the kingdom's enemies (the Russians and the Northern Coalition,) overthrow the government, exile the king, and paint new historical scrolls in which The King gets peed on by a dog.

SirMolle was in exile. Goons now stood upon the battlements of Fortress Delve, the newly captured home of a coalition that numbered tens of thousands of players. To mark the new era, a Goon artist under the name "Poluketes" truly did paint a massive 22-page papyrus-like relief depicting the Great War (excerpt on p35) in which SirMolle is unseated from his throne atop a great war elephant (symbolizing his overpowered Titan fleet,) and yes, is even peed on by a dog.

Center: A Territorial Claim Unit in Goonswarm space bearing its mascot and logo "Fatbee."

The 27 Doomsdays were a punctuation mark at the end of the Great War, the harbinger of a new post-BoB era. But if there was one thing the Jester hadn't counted on, it was that this absurdly ambitious plot would actually succeed. When it was all over, the Jester had no jokes left worth telling, and RedSwarm Federation faced an existential crisis.

RSF

The coalition known as RedSwarm Federation consisted primarily of three factions: the Russians of Red Alliance, the largely American Goons, and the French of Tau Ceti Federation.

The coalition itself was never meant to last beyond the destruction of Band of Brothers. The Goons and the French had helped Red Alliance get its revenge on the Coalition of the South, and the Reds had returned the favor by helping Goons destroy BoB. It's not even clear that any of them truly believed BoB could be destroyed. BoB was literally referred to as "omnipotent" by RedSwarm at the time (a reference to the fact that a member of BoB had been caught abusing his developer powers.) So it's doubtful that anybody in RedSwarm was giving too much thought to what was going to happen after the war. For all they knew going in, there might never have been a true ending to that ancient rivalry.

But now that an end to the war had come, there was nothing keeping the three organizations together, and each began a new and independent chapter. The path that each of them took—and their enduring trans-national friendship forged in the Great War—plays a critical role in shaping the grand storyline of *EVE* for the remainder of this book.

 TAU CETI FEDERATION

Tau Ceti Federation was invited by its Great War-allies in the Northern Coalition to travel to the northern regions and become permanent members of the largest and wealthiest (and most Euro-friendly) coalition in *EVE* now that BoB was dead. They were given control of a lucrative star region called Deklein, and settled into a peaceful post-war existence surrounded by allies and easy ISK.

The Northern Coalition viewed the Great War as the ultimate bonding experience. They decreed that the old anti-Band of Brothers team were inseparable allies who would rule over nullsec as a triumvirate, providing safe gameplay for a generation of *EVE* players. Their motto was literally "Best Friends Forever" which applied mainly to its member alliances, but was also loosely applied to everyone who had fought to destroy BoB.

With its wealth growing, Tau Ceti Federation began inviting new partners to live under its protection, including the smaller combat-focused groups "Tactical Narcotics Team" and "OWN Alliance," and together they formed the Deklein Coalition, named for their collective home region. With Tau Ceti Federation at its head, this Deklein Coalition became an important subgroup serving the much larger umbrella organization of the Northern Coalition.

 RED ALLIANCE

The Russians at last returned to their homeland in the southeast. They took up their rightful residence in C-J6MT, in the region of Insmother, after three years of war. That war had begun with the famous Siege of C-J6MT on May 25, 2006, and the Russians had managed to retain control of the system throughout the conflict. The only difference was that now it was surrounded by a veritable kingdom spanning likely more than 2000 star systems. The entire eastern half of *EVE Online* was Russian, and the famous cloning station at C-J6MT was its symbolic capital: "RA PRIME."

The leaders of the loosely aligned Russian alliances—primarily the players Death, Mactep, and Nync—built a massive rental network where any corporation could pay a fee to be given access to any of their star systems for a period of time. There were people all around the world who wanted to make ISK in nullsec but weren't involved in any of the major coalitions. Red Alliance capitalized on this by allowing others to piggyback on their sovereignty, for a price. The renters from high-security space would pay a set amount based on the type of moons and asteroid belts that spawned in that star system, and they'd be given a proof-of-purchase code they could present to any Russian defense patrols they might encounter passing through the system. As long as they kept up with their payments, they were left to exploit the star system however they wished.

With this enormous amount of territory, several of the Russian alliances were able to attract what essentially amounted to tenants. Small groups of entrepreneurs would use the alliance's systems for a week or so at a time to extract as much wealth as possible from a lucrative nullsec star system, and give up a cut of their yield—be it ISK or minerals from killing NPC drones—for the privilege of renting premium space.

With this rental empire secured, the Russians had an extremely steady source of passive income that could fund whatever they wanted to build. Cloistered deep in the Drone Regions, the core alliances of the Russian community began stockpiling huge quantities of ISK and minerals for a secret project that would eventually alter the course of the nullsec story.

But the Great War had also fractured and divided the Russian community even as it spurred growth. Nobody knew yet what had caused the growing rift, but it was becoming apparent that the godfathers of the Russian community were no longer on speaking terms.

 ## GOONSWARM

The Goons ensconced themselves in Band of Brothers' former headquarters in Delve, and began a conversation about the purpose of Goon-kind in *EVE* now that its great enemy was defeated. Goons had been fighting Band of Brothers for three years. It was a mythic rivalry as great as any in the history of video gaming. Justin vs Daigo. Flash vs Jaedong. Alliance vs Na'Vi. BoB vs Goons.

These rivalries defined the games they took place in because there was something within them for the game's community to believe in. Some of these rivalries centered on national pride, personality, or preference of playstyle. The BoB vs Goons rivalry was all of that and something more. It was a rivalry the community placed its hopes in because it spoke to the central conflict of *EVE* itself: Order vs Chaos.

The rivalry was also about how these players viewed the late-2000s internet and the virtual world of *EVE Online*. Many believed it was a sandbox that existed for nothing but fun in the present moment. Many Goons viewed the internet itself as a kind of joke realm in which nothing was real, and therefore everything was inherently silly. This made them diametrically opposed to the extreme self-seriousness of Band of Brothers.

At the end of the Great War, Chaos triumphed and Order collapsed. But when Order is defeated, Chaos loses the context in which it spread.

The war itself had been an extraordinary PR opportunity for RedSwarm, but it was also exhausting on its key membership in a real way. The final invasion of Delve called each of them to contribute hundreds or thousands of hours of work over a period of years.

I use the word "work" deliberately. Though this was ostensibly a video game, much of this event involved activities that would be absurd to call "play." Fleet operations could take hours and result in nothing more exciting than shooting a heavily-armored, undefended, stationary target just to roll the invasion forward a single moon. When they were done, the Fleet Commander put a checkmark in a master Google Doc, and proceeded to the next starbase, hundreds of times. During the pitched battles *EVE* could be life-changingly exciting, but more often than not, the game was a laborious grind driven by zeal.

It is mindblowing to consider the amount of cumulative human labor involved in these campaigns. The *EVE Online* of 2009 was vast and took hours to traverse. Conquest timescales were measured in months. Planning a war campaign meant night after night of hours-long meetings balancing the priorities of easily ruffled allies. After tens of thousands of coordinated human-hours of work, the mission of RedSwarm was finally achieved, and it moved forward into an uncertain future.

The fight to control the legacy and narrative of this event is still raging to this day, and has deep implications for all who wish to grow or maintain power in modern *EVE Online*. Factions disagree not only over what happened, but who was there, who said what, and how the story is told. In the historical writings and records of these organizations you can occasionally see characters disappear from an organization's history retroactively after running afoul of the leadership. In other cases, usurpers may challenge a leader's authority by writing a new version of the history which rearranges the cast of characters. In the canonical history, the leader may be a grand war hero, but perhaps in the new history they were actually a pompous royal who sat in the back of the battle "leading from behind." Most player-created histories are full of clear malice toward certain people. In many ways, the histories themselves exist for that purpose. They are ideological statements, defining from whose perspective the organization has agreed to see the past, who were the villains, who were the heroes, and who lost the faith along the way.

After the exhaustion of the Great War, a conversation took place among the Goons about what they should do now that Band of Brothers was finally—in their view—defeated.

It represented the completion of the group's grand unifying goal, because to them SirMolle's posturing represented hubris run amok. Goonswarm's pilots were united behind the idea that BoB was too corrupt (and annoying) to be allowed to survive. The day-to-day of the campaign to destroy BoB was often boring drudge work—anathema to the culturally impish Goons—but they were willing to commit themselves to it in service to the larger, grander joke at the core of the campaign: "wouldn't it be hilarious if we managed to beat these wannabe space emperors at their own game?" They would've been the good guys if they weren't intentionally foul.

The destruction of BoB brought with it a sense of relief, because RedSwarm was nearing the edge of its endurance. At long last, this great endeavor was over and peace was at hand. The Goons thought this was the beginning of a great reign of terror in which their griefing exploits would be backed by a massive cash flow stemming from their control of the most valuable region in nullsec. It was to be a grand return to what they believed was the true state of *EVE*: anarchy.

All the factions of nullsec, in fact, looked forward to a return to normalcy, even if it wasn't clear just what that meant anymore. Tens of thousands of new players had joined *EVE* to participate in that grand spontaneous event, and for these players life in nullsec before the Great War was just a bunch of legends the old timers talked about late at night on TeamSpeak. The Great War had drawn every other major group into an alliance. BoB vs *EVE*. So now that BoB was dead, all that was left was the massive alliance. Everybody's friends now, right? Perhaps. But as we'll soon see, these bonds of friendship came with their own complex set of problems—problems that aren't unique to a virtual world. After all, Alexander wept when there were no more worlds left to conquer, didn't he? How long could a group of powerful allies stay content with the limits of their power?

Destroying the last of the old guard power structure took away the only true purpose that Goonswarm had ever known. Now that its nemesis was dead, everything started to get quiet. There was no great enemy to lampoon in propaganda posters. No dire battles to take part in. No culturally-unifying rants to read on the forums.

Goonswarm spymaster The Mittani would later say that after the war, the Goons felt like a cat pawing at a dead mouse. The Goon wiki describes the time like this:

```
"The future for Goonswarm looked bright.
Unlimited cash supplies, the best re-
gion in the game, and high off the utter
defeat of BoB at the hands of the Coa-
lition, who had all finely tuned them-
selves in order to destroy BoB. After
years of work and collaboration, they
had achieved their goal. Still, the ques-
tion remained: how long could a horde
of supercapital-capable alliances, all
having invested ludicrous amounts of
money into their capital fleets and cap-
ital production lines, remain friendly
to each other or find something to do
after the destruction of BoB? Would they
turn on themselves? Would BoB return,
becoming the pest that never goes away?
Time would tell."
— Goon history wiki
```

DOMINION

Most established alliances were willing to give peace a brief chance, because a new expansion was coming to *EVE*. Called "Dominion," this new expansion promised to change the way sovereignty was captured and held in nullsec. Previously, sovereignty was awarded to the corporation or alliance with the most starbases around the moons in an individual star system, turning invasions into a tedious, months-long grind over hundreds of starbases. CCP announced that the Dominion expansion, scheduled for release on December 1, 2009, would do away with starbase sovereignty. Rather than battles being fought over potentially dozens of starbases throughout a star system, the new Dominion mechanics centralized the focus onto one structure known as an Infrastructure Hub or iHub. Attacking alliances would attempt to hack into and shut down the iHub's defenses so that it could be destroyed and the system conquered.

This caused a great deal of uncertainty. Much about conquest in nullsec space was about to change. Fleet commanders would have to learn new tactics and protocols,

Above: A propaganda image made by a member of Goonswarm after the conquest of Delve in which an iconic photo of the fall of Berlin in World War II was modified so that the soldier hanging the Soviet banner atop the Reichstag Building is instead hanging the hand grenade logo of Goonfleet. **Below:** A spread from "The Papyrus History of the Great War," a 22-page relief created by Goon artist "Poluketes" which depicts and satirizes the events of 2008 *EVE Online*. The vast majority of the original document is unpublishable for one offensive reason or another, and the cover is signed "In the name of the prophet Karttoon," the leader of Goonswarm at the time.

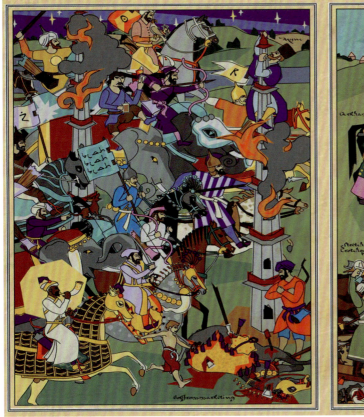

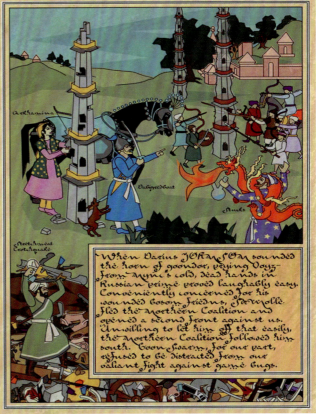

Above: A Tristan-class frigate explores an asteroid belt in the Fountain region. Opposite: The logo of Pandemic Legion.

and pilots would have to master new skills, commands, and logistical procedures. Those alliances that were the very best at starbase warfare might not be the best at iHub-based combat, at least not initially. Most groups—including the Northern Coalition and Goonswarm—wanted to avoid making bold moves right as the foundation of nullsec power was about to be shaken. In particular it seemed like the new system might cause the scale of individual battles to escalate far beyond anything seen so far.

For those who were presently out-of-power, however, this seismic shift seemed like the perfect opportunity.

IT ALLIANCE

One of the key differences between a real war and a virtual one is what's being attacked. The players themselves are never in any real danger of being killed or destroyed. The true target is the players' ability to work together. Ships, territory, bonds of friendship, and the corporate structures which make resource sharing possible—all of these had been attacked during the Great War while the players themselves remained safe behind their screens. The pilots of Band of Brothers hadn't been destroyed, they'd just been deprived of the

ability to cooperate. So what happened to its members, its fleet commanders, and perhaps most importantly, to SirMolle?

When Band of Brothers was finally destroyed, some of its people took a break from *EVE*, but many of its pilots were still in the game. In some cases they transferred to other corporations that vowed to keep hunting Goons. Others went to play in other parts of *EVE*, like the newly-introduced wormhole space (2600 uncharted, unconquerable systems that can only be accessed by travelling through semi-random wormholes.)

The founding corporations of the Band of Brothers alliance were still very much intact after the Great War. There was a dip in morale after Delve was lost to Red-Swarm, and many key members stopped playing for a while, but BoB's former leaders, SirMolle and Dianabolic were soon back at the drawing board trying to figure out how to take back their home.

They were notoriously calm people, and to them this was merely a setback. They looked at the problem as strategists, and conducted an autopsy of the lost war. After deliberation, they ultimately concluded that the reason they lost the Great War was that the game design favored Goonswarm's style of mass-pilot strategy, in which the enemy is overrun by sheer numbers. Acting on this theory, they set out on a mission to build themselves a new organization to rival the breadth of the famously populous Goons.

Though RedSwarm had destroyed BoB's ships, its defenses, and its alliance, RedSwarm couldn't destroy SirMolle's fame, notoriety, and massive social network. SirMolle and Dianabolic used those intangible assets to assemble a veritable armada of small alliances who saw this as an opportunity to get in on the ground floor of the next great *EVE* power. Many of them were the same members of the old "Greater Band of Brothers Community." But there was also a colossal diplomatic recruitment drive as tens of thousands of players were directly or indirectly lobbied to join SirMolle's new group "IT Alliance" (pronounced "it".)

"We took the valuable allies in various alliances and told them 'you are too small, form everyone into one alliance and join us on this venture," said SirMolle. "And the venture was IT Alliance. So IT Alliance was formed in one day, and went from zero to 8000 members in one day."

The data from "dotlan.net"—a player-operated website which has tracked virtually every change that has occurred in *EVE Online* since 2008 from the populations of alliances to daily soveriegnty changes and thus contains practically the entire history of *EVE* in raw numbers—shows that SirMolle's memory exaggerates a bit. IT Alliance was formed on August 18, 2009, and grew to about 4500 members—still quite large—over the coming five months. SirMolle's corporation Evolution was the first to join. A month later they were joined by groups like X13, Dark-Rising, Finfleet, Black Nova Corporation, and Reikoku. As well as renters like a young group called Nulli Secunda who would grow to significance later on. IT's mascot was a cartoon clown in front of the word "IT," a reference to the Stephen King novel. Practically everyone I spoke to besides SirMolle called them "eye-tee."

Was IT Alliance just Band of Brothers by another name? Yes and no. Some leaders in *EVE* at the time regarded IT Alliance as a wholly new entity that carried very little of BoB's old DNA. Others didn't even acknowledge the new name and kept calling them Band of Brothers. The reason for the split opinion was because while it had many of the same key membership, IT functioned off a completely different organizational ideology. Whereas Band of Brothers was like an elite military unit/country club for the old guard of *EVE*, IT Alliance would accept almost anybody who promised to participate in their campaigns at least a little.

In contrast to Band of Brothers which was run like a strict military unit, the bulk of IT Alliance was a mammoth blob of borderline strangers who SirMolle attempted to funnel into his grander revenge strategy. The DNA it retained from Band of Brothers in the top level of its leadership, namely SirMolle and Dianabolic, was enough to retain most of BoB's former enemies.

With this huge group of people IT Alliance also attracted strong allies such as the Russian splinter group Against ALL Authorities which always seemed to go against the flow of the rest of the Russian community, and traditionally held power in the south.

Former IT Alliance leaders I've spoken to told me IT Alliance was a sinister hive of ambition whose members largely joined because they were looking to suck up to SirMolle or were looking for an advantage for their corporation, rather than because of a unifying goal or culture.

With the mob of IT Alliance assembled, SirMolle began looking for the moment he would make his triumphant return to nullsec sovereignty. He was looking for a soft target to stretch out his alliance's legs and try out his new warmachine. The first effort was timed to coincide with the release of the upcoming Dominion expansion.

What is hard to capture in these pages is that it was always somewhat clear to the community that IT Alliance would ultimately go down as a mere footnote in the larger story of nullsec—widely regarded as a sort of social aftershock of the Great War. But that brief history impacted the events of this story in significant ways.

IT's first encounter would be with an old enemy from the Great War: the mercenaries Pandemic Legion. ●

DOMINION

"During the last quarter, the [...] number of characters in Dreadnoughts declined considerably, from 1,125 to 767. One plausible reason might be lower interest in territorial warfare since the announcement of sovereignty system changes in Dominion. That would also explain the increased general interest in industrial ships, as corporations and alliances shift from war mode to production mode in preparation for Dominion."
— Dr. Eyjolfur Gudmundsson, CCP Games former in-house economist

IT Alliance was spearheading an initiative it was now privately calling the Dominion Offensive, and it was secretly assembling a coalition of willing forces. The plan was to work not as a united coalition, but in loose concert with one another. Rather than a single invasion, this was to be a timed attack across several fronts. IT Alliance, Against ALL Authorities, and the Russian alliances in the Drone Regions each had private goals they wanted to accomplish. The Drone Region Russians wanted to take space away from the Northern Coalition, while Against ALL Authorities wanted to take Querious from Goonswarm, and IT Alliance wanted to take Fountain from Pandemic Legion. If any of them attacked individually, all of those enemies would have banded together and supported each other. By attacking simultaneously, however they had an opportunity to keep them divided.

The best way to imagine this is by picturing a donut sliced into six pieces, red icing on one piece and blue on the next so the colors alternate. Now we have three red pieces and three blue pieces that represent the powerblocs of 2010 nullsec. If all of the red donut pieces were to focus their attack on one member of the blue donut team in an effort to destroy it, then all of the blue donut pieces would rush to their defense, and it works the same way in reverse. That is, after all, why this equilibrium was established in the first place. The Dominion Offensive sought to disrupt that equilibrium. All of the red donut pieces would attack their clockwise blue neighbor, ensuring that each blue donut piece was engaged in a private war that would prevent their enemies from supporting any of the others.

Dominion was hyped as an expansion that would alleviate lag and allow for larger fleet battles, but word was circulating back from *EVE Online*'s Test Server that the Dominion expansion was experiencing crushing lag and disconnection issues in battles as low as 250 people, extremely low on the scale of major nullsec fleet engagements. Plus, new expansions mean new mechanics, new user interfaces, and all sorts of logistical problems brought about by the changes. Alliances throughout *EVE* were forced to learn in order to maintain sovereignty, and learning while stressed and under attack was bound to lead to errors. New expansions and patches introduce logistical problems for alliances for the simple reason that it takes time for information to spread through human networks.

When Dominion launched on December 1, 2009, IT was getting settled and preparing itself to become a potential sovereignty-holding power. It was basing its operations out of Syndicate, a region famous in *EVE* history for being the original home of the Goons back in 2005. One night its fledgling base was struck by Pandemic Legion's entire capital ship fleet, and a several hour dreadnought battle ensued in the old Goon home.

"Goons called 'em up," said Slinktress, a diplomat in the nearby Northern Coalition who was keeping tabs on the situation.

Left: Bombardment rains down on an Infrastructure Hub, the new focal point of nullsec warfare after the Dominion expansion. Center: The logo of IT Alliance.

39

When the attack was repulsed, SirMolle was enraged, and decided to bump up the Dominion Offensive plan. "We said, 'fuck it, let's go take Fountain,'" he said. "So we went from that battle, reformed ships, and then took everyone into Fountain. All in the same day."

IT Alliance's long-term target was Goonswarm-held Delve, on Fountain's southern border. SirMolle wanted to use this moment of courage to rally the everyday pilots that filled the ranks of IT Alliance to overwhelm Pandemic Legion in Fountain, and hopefully train them into something resembling a proper force capable of the assault on

Delve in the process. When it was over, the plan was to take control of the moon-mining network in Fountain to secure an ongoing cash source, and use that cash to fund a steady stream of ships to use against Goonswarm on the southern border.

The invasion of Fountain was like a practice run. The rank-and-file recruits of IT Alliance simply weren't as experienced as the people SirMolle was used to commanding.

"We had an axe to grind," said SirMolle before adding a caveat. "We had five corporations that knew each other in and out, and then we had six thousand people who knew nothing. In other words, training wheels."

The invasion was intended to give these green recruits some experience in large fleet actions. Pandemic Legion was a burgeoning power, but it stood little chance in the face of IT Alliance's massive membership which taxed the shaky Dominion server everywhere it travelled.

WE DIDN'T WANT THAT REGION ANYWAY

The massive fleet under the command of SirMolle stretched out from Syndicate to attempt to capture its first nullsec territory since the fall of Delve, being careful never to bring so many people to any particular system that the server crashed.

IT Alliance began its push into Fountain and immediately gained the upper hand when Pandemic Legion's defense instantly faltered. An ally of PL's had screwed up its sovereignty bill payment (part of a new mechanic introduced in the Dominion expansion,) causing it to suddenly lose ownership of a set of systems in the middle of Fountain. This will not be the last time in this story when a failure to click "autopay" will change history.

With those systems suddenly up-for-grabs, IT Alliance quickly seized its first stars in nullsec. Pandemic Legion CEO Shamis Orzoz reportedly commented that he didn't want those star systems anyway, and that this was ironically a good thing because now that IT Alliance was moving into nullsec it could be trapped and ground into dust.

The one advantage Pandemic Legion had to counteract IT Alliance's massive numbers was its impressive supercapital fleet, one of the best in New Eden. During the years of the Great War it had managed to build or acquire more than half a dozen of them. However, the Dominion expansion had introduced significant changes to the Titans. In the past, a Titan's Doomsday weapon was capable of wiping out dozens or hundreds of smaller

IT Alliance

OUTER RING

Pandemic Legion

FOUNTAIN

SYNDICATE

SOLITUDE

DELVE

QUERIOUS

AAA

Goonswarm

PERIOD BASIS

Above: A pilot's perspective as the Battle of Y-2ANO cascaded out of control (communications windows have been blacked out.)

ships at once in a massive, battlefield-clearing area-of-effect attack. After Dominion, the Doomsday became the polar opposite: a focused, single-target attack capable of destroying almost any single ship. (Most players were happy with this change, because the previous area-of-effect Doomsday weapons rendered any ships smaller than a Dreadnought borderline useless in fleet combat.)

This change gave IT Alliance a decisive advantage, because it could field hundreds or even thousands of battleships and cruisers that were vastly smaller and less expensive than Dreadnoughts or Titans. IT needed this advantage, because its fleet was no longer what it once was. The four Titans lost by SirMolle and the thousands of dreadnoughts, carriers, and battleships lost in the defense of Delve were gone for good. IT Alliance was starting practically from scratch.

The two fleet styles clashed on January 3, 2010, a month into the invasion, in the system Y-2ANO, in the deep south of Fountain on the border of Goon territory in Delve. IT Alliance launched an attack on the main trading station in Y-2ANO under the famous Fountain nebula—a rust-colored star factory that covered the sky.

Feeling a bit uneasy about the huge force en route to its capital, Pandemic Legion informed its allies in Goonswarm about what was happening, and a chorus of fleet commanders in Goonswarm said it was their duty to help

their old Great War allies. They added that it was also a strategic imperative, since a failing Pandemic Legion would put IT right on Goonswarm's northern border. The Goons in turn rallied the Northern Coalition who also sent a fleet. Though all were tied up facing their own attacks, the defense of Y-2ANO was considered essential to keep "Band of Brothers" from taking hold in nullsec again.

And so it was at the climax of the campaign that hundreds of IT Alliance ships found themselves facing an imposing fleet spearheaded by five Pandemic Legion Titans and entire fleets of capital ships from Goonswarm and the Northern Coalition. It was the most anticipated battle of young 2010. Pandemic Legion and its Goon/NC allies fielded a fleet more than sufficient to fend off this IT offensive, but fate conspired with buggy networking code to produce an outcome that nobody expected.

For IT Alliance, the Battle of Y-2ANO was the most triumphant moment in the history of *EVE*. For Pandemic Legion, it is remembered as less of a battle than an unjust execution. For everyone else, it's an object lesson in the realities of live game design.

"Band of Brothers was already in-system with a shitload of stuff," said Goon Fleet Commander Mister Vee, who would catch hell later because he was the person who lit the "cynosural field" (jump bridge) that warped the Goon fleet into Y-2ANO. "We knew we had more [ships] and we knew we could win the fight, but we also knew that back then system nodes were extremely unstable, and jumping in would be a massive risk because sometimes everybody just black screens and they just sit there like fish in a barrel."

Pandemic Legion's Titans arrived in the system, emerging from within the silently rippling electrical spirals of a cynosural field. But as the fleet arrived in-system, its pilots never did. Most of them never even loaded the system, because the lag was too great. So while the hulking ships materialized, their pilots couldn't fly them. Because they had gotten into position first, IT Alliance wasn't nearly as badly affected, but they didn't know that yet.

They stared across the depths of space at the defending PL/Goon/NC fleet for nearly an hour without knowledge of the other side's catastrophic connection issues. After hesitating for so long, one of its scouts named Judge Reaper began to warp to the location of the PL fleet, more than five minutes away at full speed. After finally arriving at the fleet's location in deep space, Judge Reaper found most of the fleet offline. Reaper quickly set up fleet-wide warp disruption bubbles to prevent the fleet from escaping (often shorthanded as "tackling") and called out in all-caps in the IT Alliance fleet chat channel the names of the enemy Titan pilots that should be targeted, "Titan KILLS. 5 TitanS TACKLED. MORKHT DRACK IS DOWN. NIGHT JACKAL IS MELTING. JANKO IS LOCKABLE."

Once the fleet arrived, the IT Alliance chat channel was abuzz with players trying to lock on to different PL/Goon/NC ships and then alerting their fleet in all-caps when a ship of strategic importance could actually be locked onto in the glitchy, nearly incapacitating lag. Once the fleet was locked on, the destruction of a ship took only a short time. Over the ensuing hours, a disaster unfolded and salvagers began looting wrecks from the largest massacre in the history of *EVE Online*. Footage from the battle site shows a tangled mess of derelict hulls and scattered debris.

EVE is a live game. You only get one chance at anything, which is what makes these battles so thrilling for the people involved. A million things can go wrong, and the weight of responsibility to the allies who have entrusted you with a task lies heavy. So when an opportunity presents itself, players don't stop to ask whether their enemy has had a proper amount of time to form a defense. At the first glance of an opening, the attack begins, and it doesn't stop until every single opportunity has been taken advantage of. Only after the battle reports are written by the opposing fleet commanders do they find out that the reason they won was because something went catastrophically wrong for the enemy.

The tattered remnants of the defending fleet extracted, but the damage was already done. Four of Pandemic Legion's Titans were destroyed and the Goon/NC capital ship fleet met the same fate. Roughly a quarter of all Titans ever destroyed in the game at the time were lost that day in Y-2ANO.

"Beaten not by the invaders but by the servers," lamented Bagehi, a commentator and history writer on the *EVE* news and culture website CrossingZebras.com, about Dominion's first month of live play. "A month of black screens and lost ships—not because of mistakes or inferior tactics, but because of server performance."

By the end of December, CCP patched the netcode to eliminate the bugs that kept causing the server to struggle because of large player numbers. It didn't reverse the loss of Pandemic Legion's Titan fleet, but it meant that alliances could now reliably escalate engagements and bring in reinforcements. But as soon as an improvement to server stability was added, *EVE* alliances responded by rallying even more troops. Good nullsec fleet commanders became intimately familiar with server mechanics, and knew exactly how far they could push the limits.

The battleground of *EVE* often exists on an extremely unstable foundation. Though it was weaponized most effectively during the early stages of this "Dominion Offensive," the state and performance of the server has had an extremely heavy hand in shaping the state of the game throughout all eras of *EVE*.

After this disaster, Pandemic Legion was left deflated and discouraged, and lost all interest in defending its space. Pandemic Legion was a highly coordinated group that simply couldn't deal with the IT Alliance mob, or the roiling wake of lag that it caused.

Whether it was just or not, the slaughter in Y-2ANO left Pandemic Legion in a state of in-fighting and blame-casting. When time came to debate what the alliance should do next now that its days as a sovereignty-holding power were over, it was split.

One of its most active sub-groups was a corporation called Illuminati who believed they should head to

Above: The Gallente Administrative Outpost located in the system 6VDT-H was one of the central hubs of the Fountain region.

low-security space to fight with pirate gangs while earning money doing mercenary contracts. But the others wanted to return to being a player in the nullsec sovereignty space. The majority ruled, and Illuminati was kicked out of the group. It scarcely mattered, though, because Pandemic Legion quickly began falling apart.

At this moment, a low point for Pandemic Legion and an ascendant high for IT Alliance, the citizens of *EVE* would have found it strange indeed to know the truth: that IT Alliance's days were numbered, but a galaxy-shaping destiny was in store for the now adrift Pandemic Legion. In its own records and words, Pandemic Legion refers to this alternately as their "7 Years of Darkness Period" or their "40 Years in the Desert Period."

They would return to shape events again one day, but for now there was a far more pressing concern: the northern border of the new Goonswarm stronghold was threatened by a new SirMolle-led alliance, and there was no doubt in anyone's mind that he intended to return to claim his home in Delve.

Goonswarm's leader of intelligence gathering informed the alliance of the situation:

```
"Fountain will have fallen by tonight at
approximately 3:00 eve [standard time],
since Pandemic Legion pulled out. [...]
Fifteen hours later Molle will launch
a new campaign, which inevitably will
mean us.
```

```
Since we're already being invaded
by [Against ALL Authorities,] this is
hardly a surprising move. However, the
pressure of a possible combined [US Time
Zone] force of [Against ALL Authorities]
and IT is not to be underestimated.
    :siren: THINGS TO DO :siren:
    1. Stockpile fitted combat ships,
particularly battleships, logistics and
drakes.
    2. Make sure your assets are secure.
    3. Be on jabber and alert for broad-
casts re: enemy actions.
    4. Expect to be outnumbered heavily
by incompetent people flying terrible
ships. However, as Goonswarm itself is a
testament to, this can be very dangerous."
    — The Mittani, Goonswarm
```

Along with his allies in Against ALL Authorities and an upstart alliance called "Atlas Alliance," IT formed the core of what SirMolle was now calling his "Southern Coalition."

But unbeknownst to SirMolle, as he planned his invasion of Delve, the Sword of Damocles was already swinging closer and closer to the head of the new Goon leader in Delve, a player named Karttoon.

Goonswarm was about to fall without SirMolle having to lift a finger. ●

THE SHOULDERS OF ATLAS

"The Atlas Alliance empire seemed an unfathomable fortress, spanning several regions of the *EVE* map, with vast territories and [...] stations, supported with the equally extensive Render Base. The Atlas empire was one of the pillars of NullSec life in the south of New Eden. Like every great empire of man, the Atlas was destined to fall."
— Riverini, EVENews24.com

In *EVE Online*, every action has a reaction, and every vacuum has to be filled. When Goonswarm abandoned the South with its all-out attack on Band of Brothers in 2009, that action had a number of important effects. As we've explored over the last few chapters, the most important was obviously the Great War itself, and the eventual destruction of Band of Brothers. The second most important, however, was the creation of a power vacuum in the deep southern territories the Goons abandoned.

Nullsec territory is too valuable to stay vacant for very long. Even the most derelict nullsec star region is still more profitable than what can be earned in any other part of *EVE*. So when this space became available in the middle of the Great War, with all other major alliances distracted, the Southern Coalition's junior member—Atlas Alliance—saw this as the opportunity it had been waiting for.

The young alliance had spent the past half-year as tag-along allies of larger nullsec powers as its leaders adapted to live nullsec gameplay. By 2009 it was thriving, thanks largely to its second leader, Bobby Atlas. Bobby was a highly engaged and motivated player who had ambitions to shape Atlas into a power that was capable of leading its own coalition rather than merely being a subordinate in one—as it currently was under its more famous allies, Band of Brothers and Against ALL Authorities.

Opposite: In Greek mythology, Erebus was a primordial deity born of Chaos. Seen here, the menacing Erebus Titan shines its tiny flashlights into the eternal void. Above: The logo of Atlas Alliance.

Bobby was a tireless leader who had been through half a dozen smaller scale wars, including as part of the coalition that aided SirMolle in the disastrous MAX Damage campaign from *Volume I*. That campaign in particular had given Bobby both a taste of what it was like to be part of the most powerful coalition in nullsec, and also a stern belief that he could do the job better.

Atlas was on the rise, but its leader's ambitions required the alliance to claim nullsec territory of its own. Bobby was looking for opportunities to expand so he could one day build himself a power that could go toe-to-toe with the legends of nullsec. Goonswarm's abandonment of the south was just such an opportunity. When the Goons moved out, Atlas got ready to move in. The territory it initially claimed—a region called Omist—wasn't particularly wealthy, but it was more than enough to provide a starter home for the rapidly expanding alliance.

The history of Atlas Alliance is not well-known in the *EVE* community. Its story is treated more as an esoteric subject that only ancient veterans and students of the game's history are even aware of. That's because it's tale is frustratingly anti-climactic, and illustrates perhaps better than any other episode the precarious position *EVE* alliances occupy at all times; how fragile even a mighty civilization is when held atop a single pair of shoulders.

RED SOLAR LEGION

Another opportunity would soon shape Atlas's destiny. As with much of the geopolitical situation in *EVE* at the time, it had to do with the burgeoning Russian community.

45

Above and Opposite: Two of the drones that populate the Drone Regions patrolling the rubble near a huge station.
Below and Opposite: The logos of Legion of xXDEATHXx and Solar Fleet.

Towards the end of the Great War, the Russian community of *EVE* began to grow quickly. CCP Games had released a full Russian localization of the game in 2009, and *EVE* became far more accessible to a much wider swath of Russian gamers. When those new recruits made it into the game, fresh with enthusiasm, they found an established, Russian-speaking community in nullsec complete with Russian services, Russian fleet commanders, and Russian historical heroes like Death and Mactep.

But while the Goons and the Northern Coalition were each bonded internally by their approach to gameplay (for the Goons: spreading terror, for the NC: gaining wealth) the Russians had only one bonding factor: being Russian. It was not a terribly specific bond, and it made no allowances for individuals' morals or personality. There is no Russia in *EVE*–no national boundaries to force players from a given country or culture to play together (with the notable exception of the Chinese community which was on its own server–Serenity–running parallel to the global Tranquility server.) What held the Russian players together was not so much their desire to play together, as the way in which players of other nationalities excluded and looked down on them. A lot of the players in the Russian alliances were actually Ukrainians who were

simply lumped into the same group by other players who didn't care to learn the difference.

After the Great War, however, there was no longer an entity that they perceived threatened them specifically for being Russian (or just speaking it,) and so their bond became less existentially critical. Their unity was threatened not by external forces, but, paradoxically, by their unexpected triumph over those forces.

BEST FRIENDS FOREVER

Mactep and Death had been fighting together in *EVE* for more than four years, and both of them had earned their stripes at the Siege of C-J6MT, a bonafide which established them as unimpeachable veterans who were effectively enthroned as the fathers of the Russian nullsec community. The experience had also made them into best friends. Much of nullsec lived in fear of their storied friendship.

Because of Red Alliance's deep roots in early-2000s Russia and Eastern Europe, it also came with a certain fondness for aspects of communist governance. Red Alliance's version of communism differed significantly from the real world, however, as one would expect given the uniqueness of *EVE*. The people I've spoken to describe it more like a mafioso honor system, in which each person is expected to pursue strength and wealth, but for the ultimate purpose of kicking back a significant portion of the profits to strengthen the larger whole.

As Red Alliance grew large and stable during and after the Great War, Death and Mactep's fame was seen by many within the alliance to be at-odds with a more egalitarian Russian ideal. Death and Mactep embodied a different Russian ideal, however: the political strongman.

The high council of Red Alliance wasn't pleased that cults of personality within their alliance were beginning to outshine the average members and threatening the influence of the council itself. To hear Death tell the tale, they became envious of his and Mactep's power and popularity. Third parties say that Death and Mactep were likewise growing arrogant and increasingly untethered to the mother alliance of all Russians in *EVE*.

Mactep and Death both left Red Alliance around the same time, and departed to make their own way. And so, towards the end of the Great War, they each formed their own alliances: "Solar Fleet" and "Legion of xXDEATHXx" respectively.

Mactep, the famous fleet commander. Death, the politician and industrialist renowned for operating as many as 89 *EVE Online* accounts. Each helped the other carve out a space for their alliances in the Drone Regions,

which were famously populated with dozens of small, new alliances naive to the nature of nullsec politics.

According to Death, the way the two of them carved out their first home was by using diplomacy as a subtle knife. He told me numerous stories of how they weakened their enemies by exacerbating existing border conflicts, fanning the flames until war was unavoidable, and then standing out of the way while the less experienced Drone Region alliances fought each other. When the fighting had left them weakened and exhausted, Death and Mactep would move in to conquer huge portions of territory.

This didn't sit well with the leader of the vast and powerful Northern Coalition, Vuk Lau. The small alliances in the Drone Regions were generally allowed to exist with Vuk Lau's permission. They weren't exactly allies, but they were people who had stayed on the right side of the North, and that was supposed to come with certain privileges.

"We were a bit pissed because [they] were meant to be part of the old anti-Band of Brothers allies," said Vuk Lau. "They should have been helping us, not backstabbing us. I think some of the Northern Coalition allies still had them blue [marked as allies] even as they were invading."

Members of the *EVE* community have created timelapses of the *EVE Online* sovereignty map which shows which alliances officially owned which territories

Way back in 2004, an *EVE* player named Morning Maniac was hanging out in the 'Help' chat channel giving random players advice and solutions to the tricky *EVE Online* interface (which was notoriously buggy in *EVE*'s early days.)

Eventually Morning Maniac decided they wanted to provide players with assistance that was more substantial than could happen in a brief interaction in the Help channel. So Morning Maniac founded a unique corporation called EVE University. It was neutral to all other entities in *EVE* and sought only to enrich players' experiences in the game, show them different skills and fleet roles, and "graduate" them to larger, more successful alliances.

In 2006, EVE University became part of an alliance called The Big Blue alongside industrial power NAGA and PvP alliance Four Horsemen and conquered a small amount of nullsec space in Geminate. The alliance built two outposts while maintaining what was known as a "Not Red Don't Shoot" (NRDS) diplomatic policy which essentially meant that any allied, neutral, or unknown pilots were free to fly in its space and use its outposts. Most nullsec alliances operate under the "Not Blue Shoot It" (NBSI: shoot anything that's not specifically an ally) policy.

EVE University's nullsec dreams were crushed a year later however when it was attacked by Mercenary Coalition working for an unknown client. However, it has persisted as an institution in *EVE Online* to this day.

over time. You can watch the whole history of nullsec play out in this timelapse if you know what to look for. During this period in the Drone Regions it tells an unmistakable story as one-by-one, Legion of xXDEATHXx and Solar Fleet surrounded, divided, and conquered each resident alliance. This continued, unhindered, until it came time for the final assault.

ETHEREAL CROSSING

The final battle to conquer the Drone Regions pitted a pair of inseparable allies called Ethereal Dawn [ED] and Intrepid Crossing [IRC] against a Russian coalition in February of 2009, the same time that Band of Brothers was being disbanded and invaded by RedSwarm on the other side of the star cluster.

With the Russians occupied by the invasion of Delve, ED/IRC were able to hold their ground. However, hubris got the better of them and they launched a campaign of revenge against Red Alliance in its home in Insmother. Attacking the spiritual home of the Russians is rarely a good idea.

When Red Alliance leader Silent Dodger asked Goonswarm for help in the matter, Goonswarm's intelligence director The Mittani burned a long-dormant spy and stole ED/IRC's entire capital ship fleet. With all the dominance of a playground bully, he held it over their head and promised to give it back

if they abandoned their campaign against Red Alliance. The Mittani later revealed that the thief wasn't actually one of his agents, but a Something Awful Goon who was willing to pretend he was one in order to scare ED/IRC into thinking it was a reprisal for their attack on Red Alliance. The real heist, he said, was an unrelated theft of opportunity that he'd managed to twist to tell a more advantageous story.

With no other choice, ED/IRC acquiesced, abandoned the campaign, and agreed to a ceasefire.

This did not put an end to the conflict, however. There was now a faction growing within Red Alliance that viewed ED/IRC's attack on Insmother as a grievous offense, and wanted to wipe the two alliances out of existence.

I've never been able to get a good story for what happened at this point, but some sources say there was an attempted coup within Red Alliance as that aggressive faction tried to seize control.

In the ensuing drama, a number of corporations left Red Alliance for either Solar Fleet or Legion of xXDEATHXx, and just two days later, Red Alliance broke the ceasefire and attempted a "headshot" attack on Ethereal Dawn's headquarters at ZZ5X-M.

A "headshot" attack means invading your opponent's capital system with overwhelming force, ideally to cut off its center of commerce and organization before it has a chance to wake up to the threat. Properly executed, it can end an entire war in one demoralizing engagement.

Above: Another drone courses through the wreckage-strewn wastes of a Drone Region star system. Below: The logo of Intrepid Crossing. Opposite: The logo of Ethereal Dawn.

In this case, however, the attack failed. At ZZ5X-M, the ED/IRC defense miraculously held strong against the Red Alliance counterattack. There are no hard facts about what transpired in this battle, but to me, it's easy to imagine what happened by considering how each side viewed the battle.

Red Alliance was riven by distrust and confusion in the wake of the—apparently successful—coup. For Red Alliance, this was an operation that only half of the alliance cared about, amid a time of drama and division, after it'd already lost its two legendary founders and half its alliance membership to Legion of xXDEATHXx and Solar Fleet.

For ED/IRC, by contrast, this was the grandest, most dramatic moment of their *EVE Online* lives. To them, this was a deliberately broken ceasefire. This was the same Red Alliance who in years past had obliterated Veritas Immortalis, Chimaera Pact, and Lotka Volterra. Now that fabled juggernaut, the bear of C-J6MT, had descended upon ZZ5X-M, ED/IRC's home of more than two years.

Stories are powerful weapons in *EVE*, and in this case ED/IRC was telling an inspiring story of defiance and survival. Inside Red Alliance, however, the story was much different, and it was beginning to suggest that the once famous Red Alliance was now in serious decline. The myth of Red Alliance is one of the strongest in the colloquial mythos of *EVE*, and it is arguably more important than the actual state of the alliance at any given

time. In this case, ED/IRC fought harder than they may normally have because they believed they were fighting a nullsec legend. In reality, Red Alliance at the time was in shambling organizational shape. But that same mythos has ensured that throughout *EVE* history, Red Alliance is seldom down for very long.

The surge of enthusiasm and spirit propelled ED/IRC to a defense of their home that lasted three long months. It was a time commemorated by ED/IRC with one of the more famous propaganda videos made by the *EVE* community. The story they tell of their defense against a brutish Red Alliance is soaringly triumphant. It's got choirs and grand battle sequences filmed during the actual battles of the campaign, and it tells the story as its pilots experienced it: as a story truly worth remembering, even if you weren't there. The events that transpired in the story of ED/IRC vs Red Alliance were so intense and consuming from the individual perspective that for the average pilot there was no distinguishing between this and a war twice the size. To them, *this* was the Great War.

Empires of EVE tends to zoom out to understand the large-scale conflicts between tens of thousands of people, but at every stage of this story it's important to remember that it is being experienced by real individuals from a first-person perspective.

The ED/IRC defense held for three months. Three months of late night defenses, and five hour meetings seeking to spread the Reds ever thinner until, finally, on

Red Alliance's southern border, far away from the war with ED/IRC, disaster struck the exhausted alliance.

A new voice in the south of *EVE*–Bobby Atlas–had declared that the Red menace was to be exterminated. He had seen Red Alliance struggling to evict the smaller ED/IRC and determined that the time for an offensive against the thinly stretched Russians had come. Others in nullsec warned Bobby that attacking Insmother was an overly greedy move that risked reuniting the fractured Russians, but Bobby would have none of it.

"Atlas Alliance has launched a surprise attack on [Detorid,]" wrote a reporter on EVEOnline.com in an article dated May 30, 2009. "At the time of writing, clashes are concentrated around DG-8VJ, 77S8-E and HZFG-M solar systems. Given the size of the alliances occupying Detorid, it is still unclear whether this conflict will remain a regional skirmish or rather develop into a global war."

THE RISE OF ATLAS

In the middle of 2009, Atlas was growing stronger, and Bobby Atlas believed his chance for a meteoric rise was finally here. Atlas had its foothold in Omist, and now fate had gifted Bobby with an unusual opportunity: The Russians on his border had splintered into five factions (Red Alliance, White Noise, Red.Overlord, Solar Fleet, and Legion of xXDEATHXx) and were distracted by wars on two different fronts. What's more, as mentioned in the last chapter, his allies in IT Alliance and Against ALL Authorities were planning a coordinated offensive that would soon bog down the Russians' Goon allies. For the youngest leader in the southern regions, the opportunity was obvious.

"They want space really bad and they think it's so easy to take," said a director in Legion of xXDEATHXx who went by the name 'Saint xXDEATHXx.' "We knew that they would come."

Though the Russians put forth a typical front of confidence and bravado, behind the scenes the situation was worse than they were willing to let on. Red Alliance was a shell of its former self. Its name was synonymous with grit, determination, and greatness, but at this time it couldn't muster the strength to embody those qualities. When the Atlas Alliance fleets arrived in Red Alliance-controlled Detorid, the Reds began to fall apart.

"We went into 5 of the 6 station systems and quickly ripped down the jammer in each, one by one," Atlas Alliance Fleet Commander Banlish—who is now a Twitch streamer—told EVEOnline.com. "Capitals have been deployed almost non stop... support was buzzing around doing whatever required."

There was no magic wand The Mittani could wave to save Red Alliance this time, nor turn back the mass of players pushing forward into Red Alliance territory with all the vigor that only a young and excited alliance can gather. Detorid was crushed within a week, and Red Alliance now found itself caught between Atlas in the South and ED/IRC to the North.

The fleets of Bobby Atlas took control of the famous Cloning Facility that served as the main hub station in C-J6MT. It doesn't actually clone anything, but that's what its text description called it. The value it carried was mostly emotional. After the successful defense that had cemented their alliance three years ago, the Russians had named it "RA Prime." When Bobby Atlas arrived, it was swiftly renamed "RA DONE AND GONE."

Below from left to right: The three most high-profile alliance leaders in the Russian community: Silent Dodger (Red Alliance,) UAxDEATH (Legion of xXDEATHXx,) and Mactep (Solar Fleet.)

GUN MINING

Though Red Alliance was in decline, its successors were building a kingdom in the famously poor Drone Regions. With that vast space, Death and Mactep were creating rental fiefdoms called "Shadow of xXDEATHXx," and "Solar Wing." The vast regions were no longer a frantic pool of small alliances, but were partitioned and sold by the week to renters. The territory was claimed and defended by the mother alliances, but the star systems themselves were rarely inhabited by them unless an enemy contested its control.

Throughout 2009, the Drone Regions were agriculturalized, turned into a vast field for the harvesting of digital minerals.

The system was largely adopted because the Drone Regions were a different place than the other regions of *EVE*. You couldn't make nearly as much ISK running missions in this territory as you could elsewhere. Instead, the titular drones which populated the area would drop minerals when killed. Playing in drone space resulted in so many minerals and raw materials that players began referring to it as "gun mining," a subtle jab at CCP for failing to provide gameplay variety.

Nobody cared about the Drone Regions because nobody knew what to do with all the cheap minerals. Shipping them all the way from the Drone Regions to the marketplaces of empire space was borderline impossible without being caught and shot down or establishing unrealistically vast convoys.

However, the Russians had a plan. They allowed their renters to pay for their space with a portion of the minerals they collected. While there wasn't much an alliance could do with a small amount of cheap minerals, the Russians had figured out that an astronomical amount of minerals added up to more than anybody else had imagined. In the secrecy of the most unpopular and difficult-to-reach regions of New Eden they began to pour these minerals into a secret project to reconquer their homeland, and redefine the balance of power in nullsec for years to come.

With the arrival of a serious threat in the southeast, the Russians retreated into the safety of the Drone Regions and united the old community of alliances (Red Alliance, Solar Fleet, Legion of xXDEATHXx, White Noise and Red.Overlord) to focus on wiping out ED/IRC with their combined force to clear a new home for Red Alliance. Once it was over, they would split up again, but

Red Alliance is the mother of all Russian alliances, which meant that the Russian entities of *EVE* would usually seek to maintain a place for it in the game by whatever means necessary. The Russian community of this era fought like cousins who would instantly put aside their differences and become a family again if an outsider threatened their mutual matriarch.

But Bobby Atlas would not be denied. Atlas was a growing power, and to him this was the moment that the Russian myth was exposed. No longer would he bow to the spectre of the Russian legends. Everyone had heard of how the Russians refused to ever be defeated, and would grind out victories from months of grueling guerilla warfare rather than face defeat. Bobby would put the lie to these stories, and show *EVE* that the Russians were no different than any other group.

After securing Detorid, Bobby Atlas began an assault into the historically Russian region of Insmother. The Russians fought back throughout the Summer of 2009, but it wasn't enough. After conquering Omist, Tenerifis, Detorid, and Wicked Creek, now Insmother too was placed atop the shoulders of Atlas.

In October 2009, the conquest of Insmother was complete, and Bobby Atlas looked ahead into a blindingly bright future. He now held more territory than any other single alliance in *EVE*, including the crown jewel: C-J6MT. Bobby saw in Atlas an alliance of destiny—his destiny—and he began making plans to break away from IT Alliance and Against ALL Authorities, and to begin charting a new course for himself.

But no matter the opponent, and no matter the odds, the Russians of *EVE* can never forget: C-J6MT, Insmother is home. ●

KING KARTTOON

"So few real Goons actually play this game anymore in comparison to the number [of members] required to hold space. I'm doing what I feel is the best thing I can do right now before leaving this game. Euthanasia."
— Karttoon, CEO, Goonswarm

Shortly after the fall of Band of Brothers, at *EVE Online*'s annual fan convention "Fanfest," Goon-leader Darius JOHNSON was invited to give a speech on a panel with other *EVE* leaders.

JOHNSON arrived wearing a brown Dick Tracy trenchcoat and a 1930's-era replica pilot's cap with attached flight goggles, visibly hung over. He proceeded to give a lecture about Goons and the fall of BoB which began with a description of Goonswarm. He introduced his organization by quoting a rival who had described Goons in the past.

"Goons are a bunch of uncouth, savage, crass, foul-mouthed, vulgar, immature, racist-epithet-uttering ogres who can't seem to say one sentence or make a single joke without using the worst of the worst profanity (or referring to some body function, bodily fluid, or body part,) then giggling like a 10-year-old."

Darius JOHNSON paused for dramatic effect: "Guilty as charged," he retorted, giggling like a 10-year-old.

GOONS

Despite its trollish origins, Goonswarm is still probably the most fascinating organization that has ever arisen within *EVE Online*. Even its most vocal enemies—and they are many—would have a difficult time denying that fact.

It has survived destruction half a dozen times. It is ostensibly a troll alliance, and yet it produces propaganda that is genuinely impressive. It prided itself on its outsider

Left: The non-player-owned Blood Raider Covenant station in PR-8CA in Delve. Center: The logo and mascot of Goonswarm, Fatbee.

status despite being several years old. It was now an institution in *EVE Online*, and one that had a heavy hand in shaping the modern culture of the community. However, that outsider identity is a critical element in the group's social cohesion.

The Goons of *EVE Online* are a common type of internet user who find it fun to transgress, and to trample on norms of social behavior. They often play this game in part because they see it as a sort of masquerade ball, separate from reality, in which everyone can act as a different version of themselves. The questionable morality of their actions in this place is irrelevant to them, because they believe none of it is real. "Space pixels," they call it.

It is *EVE Online*'s open nature that attracts them, because *EVE* doesn't judge them or ban them. It allows them to behave how they choose to behave, unless the harassment strays beyond the bounds of the virtual world and into real life. Even this distinction is often hard to define. For better or worse, *EVE* allows the Goons to define a place for themselves in the universe. Many within the *EVE* community argue full-throatedly that it has been for the worse. No group in *EVE Online* history—not even BoB—has been more controversial than Goonswarm. As the writer of this history I feel uneasy about upholding that characterization—even if the Goons might welcome it—but others in the *EVE* community have felt no such reservations.

Well-known *EVE* blogger Ripard Teg offered the following description of some of Goonswarm's internal programs which were suggestively named to inspire fear in enemies.

"[They have] their 'Ministry of Love' – where the enemies of Orwell's 1984 were tortured and murdered. And they have their 'Reavers' – who in [the sci-fi TV show] Firefly rape and murder their victims and wear their skin. Before 'MiniLuv' there was 'Jihadswarm,' complete with a fat bee wearing a shemagh and wielding Russian weapons and a suicide vest... do I really need to get into the symbolism of that? For virtually their entire existence, Goonswarm has advocated scamming new players, using rental and membership scams on corporations and whole alliances, and suicide ganking. [Goonswarm-led coalitions] have invaded and conquered more regions in New Eden than anyone in *EVE's* history. [...] The Mittani boasted in his most recent 'State of the Goonion' and in his most recent fireside chat of the destruction of whole alliances at [his coalition's] hands."

Since *EVE* is designed to be a cold, dark, and harsh universe, the villain is often victorious and even occasionally applauded for the mastery of their villainy. But that does nothing to change the fact that they're the villain in this place. The rest of the *EVE* community usually still treats them as such. Goonswarm wouldn't have it any other way. To call them a villain is a form of affirmation that validates their sense of self in the game. The louder their enemies wail and complain, the happier they are, and the tighter their community of like-minded peers becomes. The success of their villainy and the increasing hostility from others both puts them in danger and draws them together for protection.

Their enemies say that the Goon philosophy took things too far, and that by focusing on the distress of the players behind the avatars the Goons had crossed an important moral boundary that separated the game world from the real world.

THE GOON IMPERIUM

In the social hadron collider of *EVE*, the breadth and complexity of the Goonswarm organization was an unparalleled achievement that would not be matched in *EVE* for many years. Goonswarm is one of the largest video gamer organizations in history. But the group was not always so powerful. Goonswarm traces its *EVE* roots back to 2005, when a couple dozen members of the SomethingAwful.com forums got together and decided to try to found a group of "Goons"—as they call themselves—to play together in *EVE*.

What makes someone a Goon is simple: pay the $10 fee ("tenbux") to begin posting on the SomethingAwful.com forums. That's it. But that simple barrier works wonders. SA Goons have nothing in common with each other and yet they share a certain cultural cohesion because they've all made that $10 commitment to associate with others fond of a certain "Anything Goes" nihilism.

In *EVE*, they are united by the single-minded pursuit of fun and humor, but they often arrive at their fun by a means most people would consider backward in the extreme. Most famously, they have a long tradition of trolling and griefing. When asked about this behavior, they often say that their goal is to make a joke of the person who has allowed a video game to become so important to them that they could be driven to rage.

EVE is not the only game that has been visited en masse by SA Goons, and denizens of many games have written about them as a virus that seeks only to destroy the host game's community, an accusation that Darius JOHNSON addressed on-stage in his speech. JOHNSON's speech concluded by coining one of the most enduring mantras of Goon-kind:

"We're not here to ruin the game," he comforted. "We're here to ruin *your* game."

The nuance is seldom appreciated by their enemies.

But as the group's membership swelled into the thousands, a shared love of griefing was no longer enough. An organizational structure has developed within Goonswarm out of necessity, and it is a structure that would seem to run counter to the Goons' own impish natures. Leadership of the Goons of *EVE Online* is given to whichever player has received the most praise on the forums. Thus the role tended to go to Goons who have long histories on the forums, extremely strong writing and debate skills, an uncanny ability to motivate thousands of rogues, and who have managed not to become the butt of too many jokes. This requires a keen insight into what motivates the membership, because one of the few things that unites most Goons is an abject disrespect for authority. They're usually just waiting for a leader to screw up so they can begin heckling.

Once a leader has successfully navigated that social gauntlet, they must then attempt to build a government. This means providing authoritarian legitimacy for a host of bureaucracies, fleet commanders, and communities that loosely coordinate 10,000+ people. Some of Goonswarm's institutions even persist from government to government (certain Jabber channels, meeting places, forums, and programs, for example) allowing them to continue building on the organization that their predecessors established.

Most other alliances in *EVE Online* are simple, discrete conglomerates of player corporations. The individual corporations come together in common interest, sign a treaty of some kind, and then the corporations cooperate toward a certain end. In the early years of *EVE* this level of cooperation was a cutting edge social idea. The players designed diplomatic systems that allowed their networks to grow to sizes previously unimagined by the game's own developer.

Groups within *EVE* quickly grew so large that they pushed up against natural human limits on social group formation. It's no coincidence the original capacity for a single corporation in *EVE* was 125 people. It's just shy of "Dunbar's Number" which is a Sociology term that describes the average number of people a person can maintain relationships with. But when early *EVE* corporations began to cooperate on a mass scale they started to complain to CCP that there were no official mechanics for alliances between corporations. So CCP built those mechanics, and integrated a system for alliances into the game. There are still no mechanics in the game for managing coalitions (and most recently, super coalitions) of alliances of corporations of players. The players are making all of this up on their own, finding new ways to structure themselves in an attempt to defy Dunbar's Number.

VIRTUAL SURVIVAL

Goonswarm realized early in its history that in order to survive long-term it needed to identify its own social weaknesses. It saw that the generally accepted alliance model was inherently weak because the corporations that formed an alliance were always going to be more loyal to their own corporation than to the umbrella alliance. Nobody much cared about the alliance as a whole because it was an imaginary construct that individuals rarely had a personal connection to.

Therefore, if 10 corporations enter into an alliance, that alliance will almost certainly eventually separate again into 10 corporations more or less the same as they were before they joined the alliance. This made that theoretical alliance extremely vulnerable to social attacks, because their enemies would recognize this fact and attempt to separate them into 10 individuals again rather than 1 whole.

Thanks to its highly developed spy network, Goonswarm often had a front row seat to the dissolution of other alliances, occasionally even including detailed logs of conversations between leadership during the collapse. Its leaders realized that player groups in *EVE* have differing levels and types of social cohesion. Social groups in *EVE* are a bit like asteroids in the real universe. Some of them look massive from afar, but if you study one closely you might find that it's actually a giant ball of loose gravel that will scatter in infinite directions

Below: A nullsec stargate festooned with the Fatbee logo.

if struck firmly. If an asteroid hits your home as a solid mass it might obliterate all life, but if it strikes as a cloud of gravel then it will create a lovely meteor shower (apologies to the astrophysicists of *EVE* for the crude analogy.)

The key to winning wars in *EVE*, Goonswarm began to realize, was to find the source of the center of gravity that keeps their enemies together and nullify it somehow. Destroy the leader's reputation. Twist the storyline. Foment dissent. And if the key to winning wars was to destroy or nullify your opponent's social gravity, then the key to not losing wars was to prevent them from doing that to you. Goonswarm's social strategy involved tying its social net closer together by creating a bureaucracy that combined the members of its many corporations into units which worked together to organize the alliance, study the enemy, and hunt for their weaknesses.

THE GREAT BUREAUCRACY

Managing an alliance of 10,000+ human individuals from many different parts of the world is a constant challenge. An alliance of strangers will absolutely collapse into in-fighting almost as soon as they're allowed to, because players begin factionalizing and defining their factions to contrast themselves to their peers.

Most alliances try to manage this problem by allowing each corporation to maintain its own identity and culture. This generally placates the membership and allows them to insulate themselves from cultures clashing with other corporations. This only works for so long. As the alliance grows larger, accidents and misunderstandings become increasingly likely, and individual corporations begin competing for resources and budget. If animosity develops between two corporations, there's no mechanism for resolving it besides a strong alliance leader telling them to knock it off. That method often works—if there's someone in charge everybody respects—but its effectiveness diminishes as grievances accumulate and resentment sets in.

Goonswarm approached the problem much differently, according to former members. Whereas most alliances kept each corporation separate and discrete, Goonswarm relentlessly integrated the members of each alliance into their grander whole. They did this by creating ongoing bureaucratic programs which pulled their membership from the alliance's constituent corporations. These programs were run by members recruited from across the disparate corporations.

Those projects tied the group together by creating a more intricate web of connections between the membership, and gave them a chance to form strong personal bonds with members of the alliance's other corporations.

In essence, Goonswarm believed that the strength of the alliance's social bond depended on the little bonds between the individual members of the coalition. This worked in concert with the alliance leader's propaganda and large-scale political aims to create a group that is not just tightly knit, but also self-knitting, as these groups created a virtuous cycle that strengthened the alliance while bonding the membership.

For instance, there was an economics wing that members could apply for which analyzed Goonswarm's economic policy and tradecraft, monitoring the markets in Jita and the regional hub systems "Hel" and "Dodixie." There was a "black ops" group which focused on secret operations to minutely target and exploit enemy mistakes. Whatever type of gameplay a Goonswarm member might enjoy, there was sure to be a sub-group devoted to it. This system is how Goonswarm also ended up with a sizeable

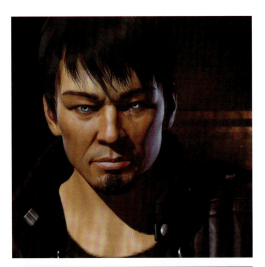

Above: An actual Goon-operated manufacturing facility photographed in nullsec in 2019. Opposite Top to Bottom: Goon fleet commander DaBigRedBoat and former Goon leader Zapawork.

Russian membership from its old coalition allies, and maintained close ties with the Russian community for years.

Members of these bureaus were drawn from all the alliance's corporations, and thus help tie those corporations together. This system keeps members focused, practised, and most importantly, it keeps them playing *EVE* in a way that is uniquely interesting to them. It gave the average member a stake in the well-being of the alliance.

Now that you know more about how Goonswarm had succeeded, you can begin to understand the circumstances that led to its incredibly rapid collapse in early 2010.

ZAPAWORK

At the peak of the Great War, the Goonswarm IRC channels and forums were buzzing with plans and discussion of the war. The forums were awash with memes and MS Paint drawings of SirMolle. But after the war passed, and the excitement had gone, membership activity in Goonswarm began to wane. Many of its senior leaders were going longer and longer without logging into the game. The Great War was an energizing event that had rallied the Goons and kept their attention on *EVE*, but what could possibly follow that act?

With the retirement of the second Goon leader Darius JOHNSON—who had guided the organization through the Great War—leadership of Goonswarm fell to a player named Zapawork.

"Zapawork was an old-school SA Goon who came from a time when Goons just [used] frigates and had fun in Syndicate and didn't own sovereignty," said Goon fleet commander DaBigRedBoat. "He changed Goonswarm in a radical way by removing a sub-forum we had called the War Room. Under his leadership he didn't believe there would be another war in Goon history. We didn't need to go to war with anyone. BoB's dead!"

By all indications, Zapawork never cared much for the drudgeries of large scale nullsec war campaigns and believed that the Great War was only necessary because BoB—by its own declaration—was an existential threat to Goons. Thus, to Zapawork, the end of the Great War was in fact an end to all Goon wars. They could at last return to their roguish calling as the great trolls of New Eden.

In addition to closing the Goon forum where it discussed war strategy, Zapawork also put a halt to military spending.

"In his eyes, there was no need for any of that stuff," said DaBigRedBoat. "There was no need for a war room, no need for strategic ops. So he was downsizing. He even told me himself that there was never going to be a need for big fleet [operations] because who needs them when there's no war, and it was going to be for the best."

Zapawork had a vision for a peacetime Goon organization. The future of Goons—as Zapawork saw it—wasn't nullsec power, but fun. That is to say, fun in the way the Goons saw it.

Much to the dismay of Goonswarm's fleet commanders, he decreed that the fleet's efforts would be focused not in nullsec, but in high security empire space, where they'd have fun by making life a living hell for average players. Pilots studied and practised the art of ganking other players (using loopholes to destroy players in secure areas before the space police can stop it) and generally wreaking havoc in dastardly creative ways.

After a few months of budget cutting, Zapawork stood down from his brief reign. The obvious choice for a successor was The Mittani, director of the Goonswarm Intelligence Agency. However, as he freely admits, his previous stint as Goonswarm director went quite poorly and so he was passed over.

With membership activity waning and alliance interest at an all-time low for the Goons, they were forced to do something they'd never done before: hand over leadership to someone who was not an old school Something Awful Goon. The leader they chose, a relative unknown named Karttoon, was an outsider to the Goons, and the first leader who didn't have a history in their motherland, the SomethingAwful.com community. He was a spy manager in the Goonswarm Intelligence Agency who had been with Goonswarm long enough that he was able to climb the ranks even though nobody actually knew him personally.

"Karttoon's passion was attacking high-sec and having fun," said DaBigRedBoat.

Karttoon was not a good peacetime leader for the Goonswarm. His contemporaries describe him as lacking a passion for organization, which is a difficult character trait to endure when your task is to run an alliance of 10,000+ people. Nevertheless, he was tasked with safeguarding the crown jewels of the Goon Kingdom, which included rare and precious possessions such as ownership of the only alliance in *EVE* named "Band of Brothers." He is said to have regularly neglected to appoint new members to top positions in the bureaucracy and many departments became understaffed.

What Karttoon and his predecessor didn't understand is that peace has a way of making alliances weak and bored, and it was causing a brain drain throughout Goonswarm as players either stopped playing *EVE*, or dispersed to play *EVE* with other groups.

DELVE IS KARTTOON'S

As Karttoon took over Goonswarm, IT Alliance defeated Pandemic Legion in Fountain and was moving in on the northern border. Remarkably, Karttoon thought little of it.

He didn't believe that IT Alliance posed any threat to the mighty Goonswarm, and steadfastly held to the decree that prevented the Goonswarm from rebuilding its wartime armada. It controlled the wealthiest region in New Eden, but it wasn't spending that money on Titans, Dreadnoughts, or building up its jump bridge network. The only thing it was being used for was wreaking havoc on newbies in high-security space.

Some of Karttoon's contemporaries pleaded with him to see the threat that was looming now that IT Alliance had managed to take over Fountain.

"I pushed hard along with a couple other guys to back Pandemic Legion up, because they were there for us [in the Great War]," said DaBigRedBoat. "Because if Fountain dies then we're next. That was the feeling. And without Pandemic Legion to back us up it would be hard to defend Delve."

However, the other prong of the Dominion Offensive, Against ALL Authorities, was attacking Goonswarm from the East. Karttoon sent DaBigRedBoat to deal with the Against ALL Authorities threat, but when he tried to form fleets he found just a few dozen Goons willing to fly on the campaign. When DaBigRedBoat inevitably failed to curtail the attack by Against ALL Authorities, Karttoon said it was a disaster, demoted DaBigRedBoat, and further entrenched himself in the belief that the hassles of nullsec were causing Goons more frustration than fun. He mused privately that it might be time for an end to come for Goonswarm.

Karttoon would later admit that he had already begun plans to destroy Goonswarm from the top down so it could start fresh.

THE SECOND FALL OF DELVE

Near the end of January 2010, Karttoon went on his honeymoon. It was a dreary time in the alliance, and by coincidence Goonswarm's Chief Financial Officer also found himself burned out on the game. With neither of them actively logging in to *EVE* they didn't notice the warning emails being sent to them by the *EVE* server that their alliance account was critically low and the auto-payment for their sovereign systems was coming due soon.

On January 26, 2010, the sovereignty bill came due and Goonswarm's corporate wallet was short 7 billion ISK to make the payment. All of its sovereignty claims simultaneously dropped, and Goonswarm's territory was returned to neutral ownership.

The heavy gates of Fortress Delve swung open and left Goonswarm vulnerable to half a decade's enemies.

While Karttoon relaxed with his betrothed in Mexico, IT Alliance fleets immediately surged toward Goonswarm space, gleefully taking advantage of the opportunity to sack the stations of their age-old rival.

```
"With the CEO absent, The Mittani called
a State of the Goonion, wherein he stat-
ed that NOL[-M9} and J-L[PX7] were both
lost, and with them the Goonswarm market
hub and Capital [ship] hangar. Without
them, it would be impossible to mount
any kind of offensive to retake Delve.
Standing orders were made to have Goons
evacuate Delve and to start making their
way to Syndicate, the cradle of [the
original goon corporation] GoonFleet.
It was declared that Delve would burn.

   In the early hours on 27 January
Junkie Beverage led a fleet to break
our capitals out of J-L station which
was already in the hands of IT Alliance.
On the 28th, evacuation ops weren't as
successful, but the 29th went smooth-
ly. Former Goonfleet CEO Darius JOHNSON
called a follow-up State of the Goonion
address for the weekend that followed."
   — Goon wiki log
```

A few days later, Karttoon finally returned from his honeymoon and found the alliance in disarray trying to salvage the situation. The actual logs from the leadership Jabber channel were preserved and we can see Karttoon logging in and getting word from alliance member Knar and former alliance leader Darius JOHNSON about what had happened. Spirits were remarkably high. Just moments prior The Mittani went AFK to get an espresso, and Vile Rat and Darius JOHNSON were joking about how American beer is bad. They were mostly laughing the whole thing off.

```
karttoon: What happened?
knar: sup karttoon
knar: we lost delve
karttoon: Yeah I saw
karttoon: How'd it happen?
```

```
Darius_johnson: Wasn't enough money in
the wallet to cover the bill
[...]
karttoon: Jesus
[...]
knar: then bob took [NOL-M9] and [J-LPX7]
karttoon: Welp
knar: trapped all our shit
knar: -welp-
knar: so how was your honeymoon
— Jabber logs from the day Goonswarm
lost sovereignty en masse
```

The general membership was far less forgiving, and the forum posts were merciless. A post was made on the internal forums polling members on whether they thought Karttoon should resign, and the result was 2:1 in favor. With their sovereignty lost, enemies at the gates, and members calling for the leader's resignation things were about to get much worse for the Goons.

THIS TREACHEROUS NIGHT

The next night at 23:00 *EVE* time, Karttoon logged in to *EVE Online*, but said nothing in Jabber. First he kicked out every corporation in the Goonswarm alliance. Next he revoked the hangar privileges of every corporation.

He also revoked everyone's access to the alliance's starbases, leaving hundreds of ships inside the shields of starbases that they didn't have permission to be inside. The *EVE Online* system has a very specific way of dealing with this permissions conflict: it fires the ship out of the station at tens of thousands of miles per second, into deep space. Goonswarm space instantly became a madcap, ridiculous scene as ships shot out of their stations at ludicrous speed the instant their pilots logged back in after downtime.

Karttoon then drained the alliance coffers of all the ISK he could get his hands on. He stole approximately 60 Dreadnoughts (worth hundreds of billions of ISK) and thousands of smaller battleships and logistics vessels. One estimate put the damage of Karttoon's theft at approximately 1 trillion ISK, a virtually unheard of sum even eight years later. At the time it could've been worth as much as €40,000.

"If it was in his access, he grabbed it all," said DaBigRedBoat. "He stole everything that wasn't nailed down, and because there were no directors active there was nobody to stop him. Since the sov was dropped, IT Alliance [took our stations immediately.] So we had to run."

On February 3rd, 2010—a month after IT Alliance's crushing victory in Y-2ANO, and almost exactly one year after Haargoth Agamar disbanded Band of Brothers—Karttoon, the very leader of Goonswarm, did essentially the same thing.

Karttoon published his official tell-all on the Something Awful forums:

"I'll try to keep this short since I've never been one for long boring stories.

I was originally planning to do this about a month ago. I prepped everything one evening with a post at hand, and moved my characters into our major cache locations. I noticed that NOL had a large logistics ship cache worth of 10's of billions in [...] ships, so I spent about two hours moving them to lowsec. By the time I was finished it was getting late, so I decided to get some sleep and mash buttons the following night. The next day I received intel that SirMolle was going to move IT alliance to work with AAA [to] invade our space, and decided to let things continue since the idea of them alarm clocking until 5AM for weeks at a time made me [giddy] inside. [...]

So I've been looking a lot over the past month at Goonswarm, and the Goons and pubbies that reside in the alliance. The majority of Goons for the most part are disinterested in eve online at this point (with good reason) [...] I've been looking at our post Dominion strategy, and strategically speaking things are easily winnable. The problem is, we'll have to pack Goonswarm with exponentially more pubbies, and negotiate standings with people I would rather never have to deal with to in order to achieve this. [...] So few real Goons actually play this game anymore in comparison to the number [of members] required to hold space. I'm doing what I feel is the best thing I can do right now before leaving this game. Euthanasia.

I've already kicked every single corp out of Goonswarm. I've also removed all of our standings with everybody. Feel free to spend the next few days blowing everyone up, or whatever. Additionally I've set GoonFleet into self destruct mode.

Tl;dr: Fuck pubbies, fuck eve, and :fuckgoons:.

Amusingly enough when I returned from a 1.5 [week] vacation to discover that most of what I was going to do was already done... by accident. I don't think I have ever laughed so hard and long in my entire life.

At this point finishing the job was obvious, but I wanted a day or two to get a few things prepared, such as unlocking all of the Titan and Mothership [Blueprint Originals] and letting the logistics crew ninja some of the stuff out for me first. [...]

I am leaving enough isk in the wallets of both executor corps to ensure the alliance bills can be paid for at least a few years [editor: he's saying it'll be impossible for them to regain access.] and I am riding into the sunset (quitting eve) leaving all the isk and assets locked in-game. I'll probably check back in a year or so and donate it all to some random goon of my choice, assuming the game still exists."

— Karttoon, Traitor, Goonswarm
Feb 4, 2010

The mightiest group in the star cluster, the champion of *EVE's* grandest struggle, was being dismantled by its own leader, and everyone was talking about it. Rumors and gossip were rife in the *EVE* community about what Karttoon had actually stolen. Legend spread that he destroyed the dreadnought fleet he stole, self-destructing them one at a time just to spite Goons (and collect on insurance payouts) though that particular legend seems to be untrue.

"Well, the job of purging Delve is going to be lighter," commented Selest Cayal, CEO of IT member corporation Nex Exercitus in an article about one of the supercapitals that was destroyed after being ejected into space by its station's shield. "The circle is completed: they disbanded BoB and now they manage to get themselves disbanded. But we also lost a great enemy. I'm sure we will see the Goons again, in some form or another."

Above: With Goonswarm sovereignty lost, IT Alliance's capital fleet set a course for Delve within hours.
Below: A propaganda image featuring the old avatar of Karttoon created by a Goonswarm member in the wake of the betrayal.

Karttoon himself confessed these crimes, but not everybody believes his version of events. Today some Goons believe all the events were connected, and that Karttoon was downsizing the alliance because he was planning to rob it all along. Other Goons believe Karttoon simply screwed up. They believe his failure as a leader came down to ineptitude, not a grand plot to destroy the alliance from within. They allege that Karttoon made up the story of a secret plot to destroy Goonswarm in order to save face, that he preferred to be reviled as the villain who orchestrated the death of the largest alliance in *EVE* rather than mocked as the buffoon whose leadership brought the storied alliance to ruin. In Goon culture, the desire to be remembered as a legendary troll rather than an incompetent dolt makes a certain amount of Goony sense.

Many in *EVE* think of Karttoon as a "puppetmaster"—diminutive slang for everyday *EVE* players who pretend to be shadow manipulators behind galactic events, but are more often writing fan fiction about themselves to cover up their own very human mundanity.

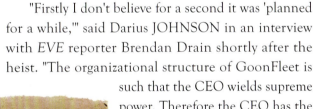

"Firstly I don't believe for a second it was 'planned for a while,'" said Darius JOHNSON in an interview with *EVE* reporter Brendan Drain shortly after the heist. "The organizational structure of GoonFleet is such that the CEO wields supreme power. Therefore the CEO has the capability to choose to or not to disband at any time. Karttoon was a terrible CEO. He made some bad decisions. Goons called him on those decisions and he chose to be a giant baby and 'punish' them by disbanding the alliance. Really there was a loss of assets, but the space had already been lost due to his lack of management ability. So the assets and [Goonfleet corporate] name were in essence just water under the bridge that goons couldn't access because he was an absentee."

DARIUS RETURNS

Karttoon may have embellished his story to make it seem like a premeditated master stroke rather than a spiteful theft of opportunity when his rule was already failing, but the result

for Goonswarm was the same. The era of Karttoon was over, and Goonswarm was left in Delve with no sovereignty, no leader, no credible defense, a severely weakened fleet, low membership participation, emboldened adversaries, and no liquid cash to begin fixing any of those problems.

When JOHNSON returned and saw the sorry state of affairs, the first item on his agenda was to get the hell out of Delve.

"The only people that could be found were The Mittani, and Darius JOHNSON comes out of nowhere and leads everyone out of Delve," said DaBigRedBoat.

He had no illusions about defending the territory he spent two years taking from Band of Brothers. He ordered that the Goons should pack up everything that hadn't been stolen by Karttoon and set a course for their old home in the star region Syndicate. They organized the core of the alliance into a corporation called "GoonWaffe" (a typically antagonistic reference to the "Luftwaffe" air force of Nazi Germany) and worked long hours to salvage the situation, trying to ensure that this devastation was not the end of Goons in *EVE*. This was, after all, the empire that JOHNSON had spent years of his life building.

One report from Darius JOHNSON suggests IT Alliance had already begun taking systems from the Goons within 10 hours of the Karttoon heist. Trapped between SirMolle, Against ALL Authorities, and its own treasonous leader, the Goon empire had collapsed completely.

"JOHNSON has stated that Goonswarm will no longer be seeking to function as a space-holding alliance, but will be pursuing goals of mischief and destruction from their new home-base in Syndicate," reads an article posted on EVEOnline.com by author Svarthol. He was taking Goonswarm back to the first home it had ever had in *EVE*.

"There is joy in the killing," Darius JOHNSON stated in the article. "There is not joy in the building."

In an alliance update he was even more direct, "As many of you may have noticed we no longer have an alliance. The very gentleman who is CEO of fleet is the likely culprit so we'll be dropping Goonfleet as well. Goonfleet is dead. Goonswarm is dead. We will now bask in the glory of Goonwaffe."

JOHNSON refashioned Goonswarm into a roving alliance, and helped stabilize the now much smaller organization. However, he was a reluctant leader. He had accepted a cybersecurity job with CCP Games and knew he wouldn't be allowed to participate in Goonswarm for very long, because CCP had changed the rules regarding developer behavior. Developers were no longer allowed to hold prominent positions in the in-game universe, a policy that also affected other players; Mercenary Coalition's leader Seleene, for example.

He wouldn't be around for long, and the Goons knew they would need to find a new leader. However, with membership draining and activity waning, nobody was left with the experience or even the desire to save the

faltering Goon alliance. Player participation was already at an all-time low before Karttoon's heist, and for many players this was a sign that—one year since the fall of BoB—the glory days of the Goons were already over.

"The year in Delve was good, but with our allies falling one by one and our enemies at the gates, it might not have lasted," reads the Goon wiki about this time. "Some goons welcomed the end of the chapter. For some, it was an opportunity to take a break, while for others, it was a return to the cradle of where the alliance launched."

With some of his final acts, Darius JOHNSON kicked out a number of smaller Goon corporations who he had always hated. "They may have been longstanding Goons, but Darius just couldn't stand them anymore so he wiped them out," said DaBigRedBoat, explaining that many of those groups eventually joined Pandemic Legion. "We were dead," he said.

Darius wasn't long for the leadership position, but before he left he put DaBigRedBoat in charge of salvaging the military situation, and also approved the creation of a diplomatic sub-division called Corps Diplomatique. "Vile Rat was in charge of that. He, Darius, and Mittens were good good good friends."

Though The Mittani and Darius JOHNSON were well-known players to the community by now, Vile Rat had existed on the fringes and in the shadows, preferring to stick to his passion—spywork and diplomacy—while Darius and The Mittani stood in the spotlight. The

player behind Vile Rat was an American IT expert named Sean Smith who worked for the US Government and was often posted at embassies around the world. His job was to ensure the internet was up-and-running at all times, and a great way to do that was by chatting constantly with people from around the world about internet spaceships.

He was a cunning player who had been scheming in online games for more than a decade. In *EVE*, he built up backchannel relations with every notable leader in nullsec. He even started a VIP Jabber channel where the disparate leaders could congregate and make deals called "Jabberlon5."

Vile Rat was an integral part of the new Goon organization given that he and The Mittani were two of the only members from the old days who were still around. Like The Mittani, he usually barely played the actual video game. Vile Rat's *EVE* took place in chat programs like Jabber where he nurtured relationships with a rogue's gallery of spies, nullsec leaders, mercenaries, and myriad other shady persons residing in *EVE*'s underbelly. The character Vile Rat was a pale man of maybe 55 years, shaved bald, with pointed eyebrows, and wrapped in the black-and-gold robes of the Amarr. He was a grim sight, and his character biography (a short section in the User Interface for players to write background info about their characters) contained only an equally grim poem called "The Second Coming" about the inevitable collapse of human civilizations which now seemed painfully prophetic. "And what rough beast," that poem asked, "its hour come round at last, slouches towards Bethlehem to be born?"

~A FUZZY MITTEN~

"Darius [JOHNSON] went and talked to our old friends Tau Ceti Federation—all the French of *EVE*—they lived in Deklein," said DaBigRedBoat who was one of the few Goons active at the time. "They invited us to come hang out and couch surf with them in Deklein. So we moved into Deklein and we started to rebuild our alliance. That's when Darius goes 'OK I've done my job Mittani you're in charge.' Darius goes back into retirement, he quits."

The leadership role passed to The Mittani, with his years of experience and deep knowledge of *EVE's* political state. As he took control of a severely weakened and demoralized Goon alliance, the forum commenters were still slinging quips about the disaster of his previous reign, but for The Mittani this was a chance to build a new legacy.

"When Mittani took over after Karttoon disbanded he kind of had a blank canvas to recreate Goons how he wanted to," said Sort Dragon, who at this time was a member of IT Alliance.

The long-time spymaster would indeed get a second chance to write a new story for himself, but for now he was left with an alliance that could barely muster 30 people in a fleet, let alone reconquer the galaxy.

The problem, he told me in an interview, was that in his previous administrations he had striven for transparency and democracy. He would listen to dissenters, engage with critics, and seek input from high ranking peers before making decisions. But after seeing how that approach had failed, he came to believe that his people didn't truly understand what was best for them. Two years later, he felt he had matured in his understanding of what it means to govern in *EVE*, and had discarded his democratic inclinations. With his next attempt at leadership, he wanted to test out some new theories about digital governance.

Above: The avatars of Vile Rat (top) and The Mittani (bottom).

"People are taught in western schools that democracy is the best form of government, but the reality is that's not true in a video game," he told me. "It can be disturbing how much people want to subsume themselves to a common cause.

"Particularly with how they respond to tyranny." ●

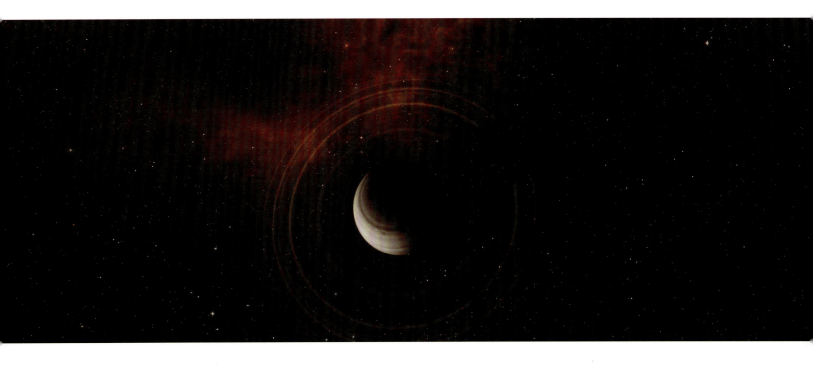

NORTH AND SOUTH

"lol good luck with that [...] a bunch of newbies aren't really
going to be able to or even know how to do anything worthwhile."
— Comment replying to the original post suggesting the creation of
a Reddit-based *EVE* corporation

In this complex and multi-layered game every decision has unintended consequences, especially decisions made at the top level of the game. These consequences ripple throughout *EVE's* galaxy, and occasionally into the real world as well. In late 2009–just a couple months before Karttoon's devastating betrayal of Goonswarm— someone posted on Reddit to ask if anyone else was interested in forming an *EVE Online* corporation. Shortly afterward, a Reddit user named Vittorios started a "subreddit" (a place for discussion of a specific topic) called r/eve.

Two months later, Karttoon's betrayal came to pass, and as the Goons collapsed in Delve, *EVE Online* was once again a media darling. Interest skyrocketed as it had many times before as the next chapter in the story of nullsec was published in news outlets across the internet. This time, however, there was a Reddit community waiting for those who Googled *EVE Online*. Previously, the *EVE* curious would have to download the game and figure out life in the cold, dark, and harsh stars for themselves. Now, they found an out-of-game community on a well-known website where they could learn the basics and find friends.

Above: A fleet of smaller battlecruisers warps past an ice-cracked world. Center: The logo of Reddit-based corporation, Dreddit.

EMPIRE STARTER

Two weeks after the Karttoon betrayal, another Reddit user—Fletcher Hammond—posted asking if he should found a Reddit corporation in *EVE*. The response was overwhelmingly enthusiastic, and one commenter suggested the name "Dreaddit." Within hours, Hammond had founded the "Dreddit" corporation. The corporation reached 125 members within an hour, and everyone was deeply annoyed when Fletcher Hammond announced that he hadn't trained the Corporation Management skill yet so everyone would have to wait 3 days for more space to open up.

Not wanting to miss this once-in-a-lifetime opportunity to ride a Reddit wave, Hammond transferred ownership of the Dreddit corporation to its second CEO, a dummy character controlled by the same player named "Empire Starter" who had the necessary skills to allow hundreds more members to flow into the corporation. Once the members had successfully joined, Empire Starter swiftly re-appointed Fletcher Hammond as CEO the next day. The corporation's by-laws were written the day after that. Many members had been witness to the historic collapses of BoB and Goonswarm, and had learned from the ordeal. The top upvoted comment on the by-laws thread reads, "Upon attaining stability the Dreddit monetary account is to be split across at least 2 different shell corporations. No single Director is to ever have access to more than one shell money corp." *EVE's* social evolution was growing ever more sophisticated, and organizations were devising new ways to inoculate themselves against rogue leaders like Haargoth Agamar and Karttoon. Dreddit even adopted an open spreadsheet accounting system which would allow anyone in the *EVE* community to see where its money was going, designed to shine light on fraud and incompetence.

On Dreddit's first day as an official corporation in New Eden, Fletcher Hammond was tasked with figuring out where the alliance would headquarter itself. He had heard in news reports that the Goons had all packed their bags and headed up North to live within the borders of the Northern Coalition in the Northwest. He decided to break ground on Dreddit's home as far away from the legendary griefers as he possibly could: in the Southeast with the 4000+ stars of Empire Space between them. It's staggering to think of all that will happen because of one simple problem: Fletcher Hammond's map was upside down.

The star cluster New Eden doesn't have a natural north or south pole, and so Fletcher Hammond inadvertently led Dreddit to a new home just 9 stargate jumps away from Goonswarm's HQ in the Northwest region of Deklein. As Dreddit headed out for one of its first mining convoy operations to gather some seed money for the natal corporation, it was descended upon by the entire Goon capital force, which crushed the mining fleet in ludicrous fashion. But what happened next surprised the awe-struck Redditors even more.

In Dreddit, Goonswarm saw a flicker of themselves long ago, before the "bitter struggle" of the Great War (as Darius JOHNSON had called it.) Bitter struggles make what is known in *EVE* as "bitter vets," a phenomenon whereby players who have seen everything become depressed, unmotivated, and often irritable. But that same bitterness gave them a certain feeling of kinship with the new and excited alliance. Dreddit's members were internet-dwellers first and foremost—just like the Goons—and their "lulz-first-ask-questions-later" attitude made them a cultural fit for the Goons, who saw a bit of their younger selves in this new alliance's origin story.

When the Goons had first arrived in New Eden they were villainized by the old guard of *EVE* as a foreign virus. "There are no goons," SirMolle had infamously declared in 2006, before BoB's initial extermination campaign. The Goons wanted to avoid the mistakes of the past, and

to offer an olive branch to the nascent group. The Goons gifted Dreddit with a hangar full of cheap beginner ships so it could get started.

Meanwhile, Dreddit's membership exploded, and the early members felt as though they were riding a surging wave of enthusiasm that was pouring new life into an old virtual world. Where the wave would crash no one yet knew, but as the first week of Dreddit's short life came to a close they had made some friends, and that was enough.

THE SOUTHERN COALITION

Meanwhile, in Delve, SirMolle and IT Alliance's core commanders congratulated themselves on bouncing back from the humiliation of the Great War. Now they would take the fight back to the Northern Coalition, who not only had fought against them in the war but had now taken in the fleeing Goons. It was a chance to rewrite another embarrassing era of BoB history: SirMolle's failed MAX Damage campaign. But this was two years after that fateful campaign, and IT Alliance was feeling stronger than ever.

"We had Titans coming out every day," said SirMolle. "We were fielding full fleets of 256 capitals. We couldn't believe it. And this was in 2010."

SirMolle planned a renewed offensive on the northern regions, called MAX 2. IT Alliance rallied its allies in the Southern Coalition, and embarked on a campaign of revenge against the northerners. The campaign was an even bigger disaster than the original MAX campaign, however, as the three main members arrived in the north weeks apart from one another and had trouble coordinating simple things like attack timings. SirMolle's new strategy of letting just anyone into the coalition seemed to be hurting IT Alliance after all.

According to IT Alliance fleet commander Sort Dragon, "the chaos set in."

The Northern Coalition's recent addition of Goonswarm and its new pal Dreddit, on the other hand, swelled the coalition's numbers beyond its already massive size. They were able to force IT Alliance back with the sheer weight of their "blob"—the demeaning name given to the barely coordinated mass of pilots who often showed up for fleet operations. Since they were defending their home territory, their lack of coordination was less of a problem than it was for the attackers.

In addition to the blob, fleet commanders in the Northern Coalition focused in on making life a living hell for SirMolle. Their ultimate strategy was to force him to retreat

Above: A large fleet composed of Rokh and Maelstrom-class battleships lands on a pink cynosural field near its command ship.

by sapping all the fun out of the game for the pilots of his fleet. Goon fleet command Mister Vee had fought SirMolle at the disaster of Y-2ANO and had no compunctions about using every tool possible to win.

"We had this Dutch guy, Porky," said Mister Vee in an interview. "He was so deeply embedded into [IT Alliance] that he was always in their fleet, and he was our main source of intel. He was so good at reading what [SirMolle was going to do] next that it gave me and our [bomber fleet] time to get setup exactly [where SirMolle was headed.] There were times when SirMolle was so afraid of our bombers that he would jump his fleet through a stargate one letter at a time. Like, literally, 'if your name starts with an A, jump now, and then the B's went, and the C's went. [Editor: he's describing a process of SirMolle spreading out the fleet so they were never a ripe target for bombers as they would be if they all jumped through a stargate at the same time.] So what would usually be a 10 minute travel time for them would be stretched up to an hour. They were completely demoralized by the end of it."

Mister Vee was able to keep tabs on SirMolle's mental state because he met with Porky regularly to discuss which of his tactics had most effectively gotten under the commander's skin.

"Tribute is falling and it came down to the staging system, the headshot attempt by IT Alliance on H-W which was the home of Morsus Mihi, the place the entire North had moved into to defend it," said DaBigRedBoat. "Somehow, someway we pulled off a miracle."

"When H-W[9TY] was threatened, it was like walking into a bear cave in the middle of winter and popping off a few rounds for lols then sticking around to see if you get eaten or not," said Bobby Atlas who was flying under SirMolle's command. "Well ladies, we got eaten."

When the invading forces made their final attempt to capture H-W9TY—the capital of Vuk Lau's home region of Tribute—the Northern Coalition's enormous membership overwhelmed the attackers and forced a retreat. "Fortress Tribute" held secure once again. One loss led to the next, and the next, as the Northern Coalition's blob overwhelmed both the IT fleets and the Dominion-era server. The second MAX campaign became a mirror of the first, a quagmire that threatened the nascent coalition's confidence. The Goons pecked away at SirMolle's patience while Deklein Coalition members like Circle-of-Two and their infamously fiery fleet commander "Gigx" (pron. "gigs") pushed the invaders back out of the north.

"The problem with IT Alliance is that we never got a hard win," said one of IT Alliance's main fleet commanders, Sort Dragon. "Too many easy wins. The Goons disbanded so we won Delve. We got the massive disconnect in Y-2[A-NO, against Pandemic Legion.] When we tried the MAX campaign we got our shit kicked in because it was an enemy that actually fought back."

As the young Southern Coalition's coordination gradually dissolved it was left with lingering distrust and disrespect between the leadership. Each was certain it was the others' fault. Respect began to dwindle, and egos began to grow. Bobby Atlas, especially, was beginning to believe the disastrous campaign was evidence that SirMolle had lost a step. ●

A COUCH IN DEKLEIN

"H-W9TY [in the region of Tribute,] stronghold capital system of the Northern Coalition, was nullsec's version of
Jita. The Northern Coalition's territories and [ally] list
spanned most of the northern half of New Eden. Tribute was a
beacon of civilization in an otherwise barbaric galaxy."
— Vik Reddy, Gamer Tribute

There is a sense of permanence that certain groups and people can obtain in *EVE Online* when they've been in the game a long time. For players going about their lives in-game these people and their power structures can feel like an immutable fact of life. Organizations persist for years, until the individuals inside them can no longer imagine life in *EVE* without them. The only true constant in *EVE*, however, is the very change that is so impossible to imagine.

The older and more sophisticated members of the community know the truth: that nothing is eternal and that claims to power are illusory and temporary. But not all players have the benefit of experience. Newbies enter into an online universe that is old and full of people who are massively wealthier and more powerful in terms of income, firepower, social savvy, and game knowledge. They are told a history that details why the power structure is the way it is. For a new player to reject the established power structure and attempt to institute their own would be like a child attempting world conquest.

As a result, from the perspective of many players, their leaders can become legends whose position of power and privilege is unquestioned. The longer some-one or some-alliance stays at the top, the more players there are who never knew life in *EVE* before them, which only enhances the mystique of their position.

If you were to have asked any player in 2010 who was the most powerful person in *EVE*, their answer would have been Vuk Lau. In *EVE*, he was the leader of the Northern Coalition, commander of the largest and wealthiest fleet in New Eden. In real life, the owner of a net cafe in Belgrade, Serbia.

The player groups who run *EVE* become far more real than the fictional factions of the game's lore. If you wanted to accomplish something in 2010 *EVE*, the person you needed to talk to was Vuk Lau, not High Priestess Jamyl Sarum of the Amarr Empire.

With that said, nobody I've ever spoken to has been able to explain why Vuk Lau was the leader of the Northern Coalition, but his success was arguably greater than any other individual in the history of *EVE*. Many in his time alleged that this success led to real world gains as well, supposedly, to the tune of millions of Euros from an elaborate ISK-selling operation.

I had the chance to interview him once, and he denied it. He talked about his time in the game as though it was a calling. He described to me his experience of applying to join a corporation which was a member of the Eastern European alliance Morsus Mihi, which was one of the two main pillars of the Northern Coalition.

He told me he met with a recruiter who gave him an application to fill out which asked questions about his history in *EVE* and his ambition for the future. The final question asked, "What do you hope to achieve in *EVE Online?*" To which Lau said he wrote: "To be a great leader."

Opposite: The Sanctuary Fullerene Loom station near the 14th planet in the Thera system.

NORTHERN COALITION

After years of squabbling and in-fighting in the early years of *EVE* what eventually brought the alliances of the north together was Band of Brothers. Or, perhaps more accurately, fear of the specter of old BoB.

The result was a group that came together for mutual defense and was bonded by external aggression. That external aggression created a central villain—SirMolle—for tens of thousands of people to unite in opposition to. The Northern Coalition was a counterbalance in the community created to offset SirMolle's blatant despotism and aggression.

At the time, SirMolle was quoted in the New York Times as saying he intended to conquer all of *EVE*, and his demeanor matched that dark goal. The alliances of the North formed a defensive pact, vowing that the northern regions were a distinct area of *EVE* and that the protection of it from Band of Brothers was a grander goal than any of their individual squabbles.

"We were all joined across the common idea that BoB represented everything wrong in this game," Vuk Lau told me. "We were holding a lot of grudge against BoB. It's a fact, BoB was abusing every possible game mechanic, and back then CCP developers were allowed

to play in [nullsec] alliances so they really had an unfair upper hand. And SirMolle was playing perfectly his role as archenemy of everything good. I hated the guy. Not even just in-game but personally for a long time. Until I eventually realized that actually he is a good guy. But back then there was a lot of emotion in the game. I couldn't believe that any game could be so immersive as *EVE Online*."

Then Band of Brothers actually invaded the Northern Coalition with its 2008 MAX Damage campaign. The campaign was designed specifically to incapacitate the north's ability to cooperate and wage war. To SirMolle the campaign was like a fun hunting trip with his members. But to the Northern Coalition it was the fulfillment of a prophecy—that one day BoB would come to destroy them.

The animosity continued throughout the Great War and beyond, and the experience of defying SirMolle's MAX campaigns was quite obviously thrilling for the Northern Coalition. Vuk Lau was full of pride that he had made good on his organization's mission statement to defend itself against Band of Brothers. The surge of morale propelled the coalition into a golden age.

And through all the threats it faced—BoB, Southerners, Russians—The Northern Coalition had beaten just about everyone and survived just about everything. Though it was derided by much of the existing *EVE* community as an alliance of "carebears"—a diminutive word used by *EVE* players to describe a player who is scared to engage in combat—the Northern Coalition embraced this image. Its art and propaganda was filled with sparkly rainbows and, well, actual Carebears.

The Northern Coalition adopted an alliance motto, usually typed in all-caps and surrounded by cartoon hearts and colorful bears:

"<3BEST FRIENDS FOREVER!!!<3"

Despite its reputation for supposed incompetence, the NC was a marvel of stability. From its perch at the top of *EVE's* map it watched the mightiest empires of *EVE's* history rise and fall. And yet for years it managed to remain steady, enjoying more or less the same borders and core member alliances.

That core consisted of two of the oldest European alliances in the game: RAZOR and Morsus Mihi. RAZOR had held the northernmost region in *EVE* (Tenal) since 2004 after the Great Northern War. Morsus Mihi was a newer group composed primarily of players from the Balkans and Serbia, including Vuk Lau and many other players from Belgrade.

The bonds of friendship within these alliances contributed greatly to their survival. Vuk tells of one occasion when much of Morsus Mihi's leadership had gathered in Belgrade for a party only to find out that one of their stations was under attack in-game by an old rival who had found out they were all AFK. When the on-duty fleet commander texted Vuk Lau the red alert they lamented that much of the alliance's core personnel were offline at the gathering. "If only you guys had computers..." the fleet commander mused.

"No problem," Lau replied. He drove everybody to his netcafe, and twenty minutes later they defended the station on LAN. Best friends forever.

H-W9TY

The coalition itself was designed as a triumvirate that divided power and wealth among three "full member" alliances: Morsus Mihi, RAZOR, and a third member which usually changed about every six months.

At the time of IT Alliance's invasion, the third member was—and had been for an unusually long period of time—Tau Ceti Federation. TCF was a French Alliance which had put the "Federation" in "RedSwarm Federation" a year earlier, and in the wake of the Great War had approached their European brethren about putting down roots in the North. *EVE* Europeans don't always share an intercontinental bond with other Europeans, but the French were accepted as a member because they would bring activity to the North during its peak European hours. This, combined with the implicit cultural diversity, made the North a busier, more active, and more interesting place to play *EVE* for every alliance's members.

In the charter of the Northern Coalition, it is stipulated that each of the three full member alliances are allowed to host three "guest" alliances within their borders. Those alliances—or "pets" as the community calls them—were granted the right to engage in commerce and gameplay in the North, and were also given a small number of star systems to take sovereignty over.

The result was that the Northern Coalition capital in H-W9TY (Tribute) at this time was a grand hub of people and languages in *EVE*. There were more than a dozen alliances, primarily from around Northern and Eastern Europe, bringing wealth and unprecedented prosperity to the North. It was nearly as bustling as the trading hub in Jita, but entirely player-conceived, player-built, and player-operated. In 2010, H-W9TY was the jewel of nullsec. It sat nearly at the center of an elaborate network of jump bridges constructed by the coalition to bypass the traditional stargate routes, in part to make industry more efficient and in part to help them reach their front lines quickly in case their fabled enemy returned.

Shipyards operated by players in hundreds of systems across the North churned out gear, ships, and equipment which would then be shipped out for sale in regional marketplaces all across New Eden, to be purchased by the vastest playerbase *EVE Online* had ever known.

To utilize this wealth the Northern Coalition instituted *EVE's* largest "Ship Replacement Program." Because destruction is permanent in *EVE*, player groups often ran into a problem that is difficult to solve: how do you encourage your members to use their ships and put their assets at risk when the safer strategy is to leave them docked in a station? If your members aren't flying their ships then it's likely their skills will start to atrophy, and worse yet, nobody will be having fun. Nullsec leaders didn't want their pilots to face a crisis of loyalty every time there was a battle because they didn't want to put their best ships on the front line.

Their solution was the SRP (Ship Replacement Program.) An SRP was effectively a promise from an alliance to its players that they would reimburse them a set amount of money if their ship was destroyed while conducting alliance business. The one stipulation was that in order to be eligible for a payout, the ship needed to be equipped according to specific alliance guidelines, because nobody wanted the alliance to go broke reimbursing pilots who weren't following the latest fleet design theories. In effect it is a form of large-scale training program that economically incentivizes players to stay up-to-date on the latest instructions.

This was a way that nullsec alliances could use their wealth to actually provide an enhanced gameplay experience for players. It was effectively a social safety net established in order to encourage positive gameplay practices. This had the three-fold effect of 1) encouraging people to use their ships more, 2) educating the average member about fleet strategies, and 3) enhancing their fleet-readiness by giving people more experience in fleet engagements. Many alliances instituted such programs, but the Northern Coalition's was by far the most generous, due in large part to the organization's vast wealth.

The wealth of the Northern Coalition—both in terms of its huge membership and its deep pockets—was aided by the surge in subscriptions to *EVE Online* at this time. Word had reached the outside world that something fascinating was happening in *EVE*, and the number of people exploring the game rose steadily between 2003 and 2010.

In contrast to most massively multiplayer online games, which tend to grow extremely large in their first two years and then steadily lose subscribers over time, *EVE Online* was a tougher sell and it had taken years for it to establish itself as a truly popular game. *EVE* retains players at a famously poor rate. The experience of playing the game itself is often much slower than video gamers are used to, and the experience of playing it alone is derided by its own players as likely the worst of all major online games. The majority of people who try *EVE* will quit almost immediately. However, a small percentage of people manage to get through that, learn to read the game's complex language, and find a community. And when they do, they tend to stay for an extremely long time. Seven years into its lifespan, this slow, deliberate pace was beginning to pay off. The vision CCP Games had for *EVE* was expanding as well. The company had now been working in secret for years on a new expansion that would change the *EVE* universe forever by allowing players to dock their ships in a station and walk around as their avatars. With *EVE* succeeding beyond CCP's wildest dreams, they decided to dream bigger.

By 2010, more than 350,000 people were subscribed to the game, and almost a seventh of them were in the Northern Coalition in one form or another, selling goods en masse to the other six sevenths. Those customers included a surprising number of eager Russians hailing from the Drone Regions, many of whom were seeking a grey market for nullsec's most tightly controlled strategic asset: Titans.

MANY MOONS

What changed the Northern Coalition from a regional power into the dominant power of *EVE Online* was a resource called technetium.

For most of *EVE's* history, technetium was "an obscure resource of moderate value" (according to the *EVE* wiki) gathered exclusively from moon-mining operations in nullsec. Moon-mining was how most nullsec alliances made a nice living, and afforded the expensive fleets used in war campaigns. A mining array would be setup modularly onto a starbase that orbited the moon, and would begin passively collecting mining yield from the surface below over time.

When the Dominion expansion was released it introduced a whole new type of equipment and ships called "Tech 2." These were leaps and bounds better than their older predecessors in the Tech 1 class, and were absolutely critical for anyone looking to do just about anything at the top level of *EVE*.

However, there was a bottleneck resource in the production of Tech 2 items. All things Tech 2 required technetium. This once-obscure mineral was suddenly at the center of an economy that would shape the next few years of gameplay in *EVE*.

Technetium could only be collected by mining a specific type of moon called an "R64" moon. These moons were previously about as valuable as any other commonplace moon found anywhere in nullsec, but the introduction of Tech 2 items created a massive new demand for technetium from players throughout *EVE Online* who needed it in order to construct the equipment necessary to compete at the top level of gameplay.

By a quirk of the topography of *EVE*, the vast majority of technetium-producing moons were located in the northern regions, within the borders of Northern Coalition members. The forums exploded with accusations of developer misconduct unfairly favoring a segment of players yet again.

More than 200 technetium moons were under Northern Coalition control. This meant that the resource itself was actually quite plentiful. There was no shortage of the material being mined, so it wasn't valuable because of natural scarcity. But market enthusiasts within the Northern Coalition realized that they could constrict the supply and artificially inflate the price.

The Northern Coalition was already the oldest, most established, and most militarily dominant power in *EVE Online*, controlling a third of the map. Then Dominion came, and it was like the Northern Coalition had suddenly discovered a massive oil deposit right under its feet.

The resources of these moons were used to further consolidate the Northern Coalition's position of power. Technetium mining operations were gifted to the most loyal people and corporations, who then gained a powerful incentive for falling in line with the mother alliance. The stakeholders in the coalition essentially received a massive, ongoing payout in the form of this new Technetium cartel.

The result was a massive period of growth for the coalition, as they could now fund any type of gameplay that anyone could dream up. The two most obvious signs of the Northern Coalition's wealth at this time was its enormous membership (50,000 characters at its height) and its unrivalled production of supercapital ships (supercarriers and Titans.) Its ability to produce ships became so vast, regular, and safe that even as the Northern Coalition built a massive reserve of dozens of Titans for its own military, some of its industrialists were openly building Titans to sell on the market in Jita. Thanks to these enterprising industrialists, any

sufficiently rich person could now buy their own supercapital ship, an honor previously reserved for only the most powerful nullsec fleet commanders. So stable was the Northern Coalition's place at the top of the *EVE* food chain that its corporations were selling off the most valuable strategic tools in the game to anyone with cash, secure in the knowledge that the Northern Coalition's massive membership made the alliance essentially untouchable.

Dominion was, after all, a notoriously unstable game build, and that instability worked to the Northern Coalition's advantage. It was infamous for using its throngs of pilots to its fullest advantage to "blob" a system with ships and either overwhelm the enemy with superior numbers, or overwhelm the server and end the attack that way. Crashing the server on purpose was against the terms of service... but if it happened to crash in the normal course of escalating a battle to gain a tactical advantage... well then one could hardly be blamed.

"They had so many members," said Manfred Sideous, at this time the main fleet commander of rival southern alliance Against ALL Authorities. "Their

Below: The steadily rising per unit price of Technetium nearly doubled from January to April 2010.

whole identity was based on safety through numbers so they'd recruit as many members as possible. There was a lot of hate for them, because they'd say, 'We'll come blob you to death,' and they did that a number of times. They'd just shove thousands upon thousands of people into a system...it was like clogging your guns with bodies."

It was through these tactics—overwhelming numbers, and a willingness to rally those numbers to the mutual defense of the coalition—that the Northern Coalition became the most powerful group in *EVE Online* history. One year since the end of the Great War—amid the collapse of Red Alliance and Goonswarm—the Northern Coalition BFFs were the only old power still remaining, and they had inherited the wealth of the star cluster.

In March 2010, near the zenith of the Northern Coalition's power, Vuk Lau was informed that one of his full member corporations, Tau Ceti Federation, had invited a new group to become its "pet" in the region of Outer Ring. The new pet was an alliance nobody had ever heard of, called "SOLODRAKBANSOLO-DRAKBANSO." It was Goonswarm, flying under a joke name they made up after the betrayal of Karttoon forced them to create a new alliance.

OPENING THE JAILS

The name "SOLODRAKBANSOLODRAKBANSO" was a Goon in-joke after they realized that in an alliance of scammers and scoundrels there was nobody they could trust. Nobody, they reasoned, except their server admin Solo Drakban, who had never betrayed them despite having complete access to all of their forums and private communications since their beginning. It was an all-caps, character-limited prayer that they be spared another betrayal at this vulnerable moment.

Once the Goons (nee SOLODRAKBANSOLO-DRANKBANSO) got resettled under their new leader, The Mittani, they refounded the organization as "Goonswarm Federation," and though the prior name was forgotten, it nonetheless does a great job of illustrating the Goon mindset at this time.

This was not the same Goonswarm that had destroyed Lotka Volterra and Band of Brothers. Nor was it the grand war machine that had sounded the 27 Doomsday cannons in 49-U6U. Goonswarm had undergone a cataclysmic metamorphoses and arrived in the north sputtering, low on fuel, chased by the IT Alliance mob, and with members abandoning ship at an astounding rate.

But at the very least, Goons still had some friends. The bond of the old RedSwarm Federation was still strong, and the Northern Coalition's "anti-BoB BFFs" diplomacy was still in effect. Tau Ceti Federation had offered to take the Goons under its wing in its home region of Deklein, within the borders of the Northern Coalition. The *EVE* community jeered that the Goons were convoying to Deklein "to crash on TCF's couch." It was on that couch that The Mittani plotted to rebuild all that Goons had lost, and bring the alliance back to its former infamy.

The Mittani arrived in Outer Ring with a Goon organization that was a shell of its former self. In an interview, he described Goonswarm as barely capable of getting 30 pilots into a fleet. Most Goons from the Great War generation had had their fun and moved on; to other games, and to other parts of their lives. Many of those who left still treasured their identity as Goons, but they didn't care to be a part of this era of the alliance in which they were forced to engage in tough, time-consuming parts of *EVE's* gameplay rather than indulging in their singular joy: lulz.

The first thing The Mittani needed to do was to raise participation back up in the alliance, but this was no easy task because Goonswarm was exclusive about membership. To become a member you had to be a Goon, which simply meant you had to pay the $10 fee for posting on SomethingAwful.com. But with only Goons to recruit from, and with most of the Goons moved on from *EVE*, the famously populous alliance had lost its eponymous "swarm."

The Mittani could be a bit theatrical, even in real life. He liked to use historical parallels of grand drama to describe *EVE* because, though perhaps a bit hyperbolic, his analogies have a way of helping people cut through the game mechanics and understand the root story behind what he's trying to communicate. So I had a feeling I knew what he meant when—in describing how he brought Goonswarm back from the brink of complete collapse—he told me that he "opened the jails."

Everyone who had previously been banned from Goonswarm was allowed back in. Hundreds of players who had formerly been deemed too vile, too disruptive, or simply too annoying to be allowed in Goonswarm at its height, were given a second chance to prove they could be decent Goons, ("decent" according to their own proprietary moral compass.) To raise their combat readiness, the Goons picked fights with small corporations and used low-stakes battles to begin training up a whole new generation of fleet commanders and pilots.

Maybe it was the difference in governance, or maybe it was that their fleets were so often filled with the formerly-banned, but Goonswarm never quite meshed with the established order of the north.

Goonswarm has always been an autocracy ruled by a single leader. The Northern Coalition, by contrast, was dominated by council governments which settled most decisions with votes from the corporation CEOs. In day-to-day business, the buck stopped with Vuk, but for alliance-level actions, the council had to have its say.

"The Northern Coalition and Goon 'culture' was very different historically and it came up every time we deployed with them or helped them out anywhere or had anything to do with them at all," said Vuk Lau.

Suffice to say, the Goons and the rest of the Northern Coalition—while united in diplomacy—were decidedly not BFFs. The Mittani, for his part, agrees with this analysis.

"There was a massive culture clash," The Mittani told me. "Some of [the Northern Coalition] despised us. Others were afraid of us."

MITTANIGRAD

When the Goons first arrived in Deklein they thought they would be under the wing of the more stable Tau Ceti Federation, but they quickly discovered that TCF was no longer the alliance it once was during the Great War.

TCF was struggling with its own organizational problems after four years in the game. The peace that was supposed to be its just reward in the north became its greatest challenge, as many of its combat-focused corporations left the alliance looking for a new fight. In response, TCF had to rely on its guests to fill its fleets and secure itself as a power. TCF itself was merely the nominal leader of a group of alliances more accurately referred to as the Deklein Coalition: Tactical Narcotics Team, Defi4nt, BCA, and a unique pvp-focused alliance called OWN which had a habit of ignoring its Tau Ceti overlords.

As the Goons arrived in the north, the leaders of the French community who had formed TCF "to unite the passion of the French people and to forge it into common objectives and activities" had come to the conclusion that there were serious problems in the alliance.

The main leaders of TCF got together for a meeting to decide the fate of the alliance, and discussed all of the issues: TCF member activity was on the decline, Defi4nt was—true to its name—refusing to pay back debts to TCF mostly because TCF no longer had the strength to force it to, and to make matters worse Tactical Narcotics Team hated Goonswarm and the leader of the corporation harbored a deep grudge against a Goon subgroup. Every alliance in *EVE* is a tangled mess of power trips, informal agreements, and grudges, but Tau Ceti's 'Deklein Coalition' was in worse shape than most.

One day a Tau Ceti council member named Aranthil made a long-awaited announcement. In the following quote I took the liberty of cleaning up some of Aranthil's English because his statement deserves to be read with the sincerity its author intended:

"We [have been] decaying [for] a few months, becoming a shadow of our former self," Aranthil wrote in a message to the entire alliance. "Yesterday the leadership council decided this should end."

"A majority of TCF's previous corporations have now joined Goonswarm Federation, a member of the Northern Coalition. We want to keep the memories of every [heart-pounding] moment, of every breath we had together in this long adventure with our friends."

And so, after four years, Tau Ceti Federation packed up its hangars, and began organizing an orderly shutdown. The goods and ships were divided up between the corporations, and the largest French-exclusive organization of players was allowed to perish peaceably. But the question then remained: with Tau Ceti gone, who owns Deklein?

"TCF starts having internal drama, and they decide they want to go their separate ways so TCF dissolves," said Goon fleet commander DaBigRedBoat. "So they gave all their space to us. They gave everything to us. And that upset a lot of people in the north. A whole lot of people. Not too many people were happy that Goons were given everything."

The main station in the capital system of Deklein "VFK-IV" was re-named "VFKhaaaaaaaaaaaaan!" for a while before settling on a more permanent title: "Mittanigrad."

Goonswarm, for its part, was a victor in this situation in several ways. First of all, it was the landing pad for most of Tau Ceti Federation's remaining corporations, which meant an influx of hundreds of dedicated and experienced pilots who were now free to staff the Goon bureaucracy and fly in its fleets without the baggage of its former alliance. Second, Tau Ceti Federation bequeathed all of its star systems in Deklein to Goonswarm. And most crucially of all, it also inherited TCF's spot as one of the three "full member" alliances in the Northern Coalition itself, and thus became the official leader of the Deklein Coalition.

This tragic moment for the French community was an enormous boon for the struggling Goonswarm, who used this as an opportunity to solidify its position in nullsec. With the handover of Tau Ceti's systems it now had a consistent cashflow in the form of the region's technetium moons. This gave Goonswarm not only financial security but, in geopolitical terms, a seat of power.

However, the diplomatic situation with the other members of the Deklein Coalition was far from tidy. They had been part of the coalition far longer than Goonswarm, and many were displeased that they hadn't been treated preferentially. These other members, Tau Ceti Federation's "guests," had been allowed to stay in the north under TCF's watch, but The Mittani had no connection to these alliances, and often saw no reason to honor previous agreements with trouble-making groups.

"When TCF announced that it would be pulling out of Deklein and handing stewardship of the region to Goonswarm, chaos erupted among the Deklein Coalition," he wrote in a memoir about this time. "Defi4nt, which had previously been able to ignore its massive financial debts to TCF, was evicted for failure to pay. The leader of TNT, who held a long-standing and curious grudge against a small corp within Goonswarm, left his alliance and began shooting Goons."

The former Deklein Coalition fell apart rapidly as The Mittani evicted those who wouldn't fall in line with the new order. The remaining members began to coalesce behind a new Goon-led coalition. It kept the same name, and it still counted many of the same

Goon and Tau Ceti pilots among its membership, but the leadership was all Goon. For The Mittani this represented a tremendous diplomatic victory which helped Goonswarm stabilize during a critical moment. Though he was ostensibly still under the command of Vuk Lau and the Northern Coalition high council, some within the Northern Coalition murmured warnings that The Mittani's latest move had effectively split the Northern Coalition into two roughly equal power centers. Others added that The Mittani was not the type to share power with a democratic council.

One of the first things Goonswarm did was invite some of its own pets to occupy the space of those who had been evicted in The Mittani's purge. In particular, there was one up-and-coming alliance who had caught the eye and warmed the heart of the normally quite nihilistic Goons—Dreddit.

Reddit's presence was continuing to grow in New Eden. The Goons loved Dreddit's style, and its throngs of new pilots had proved a crude-but-effective tool for shutting down SirMolle's ill-fated "MAX 2" revenge campaign. Goons wanted to help nurture this nascent group of young pilots who they suspected could one day grow strong enough to help Goonswarm reshape the established order. But whether that relationship would endure as Dreddit grew ever more experienced and bold was anybody's guess.

Dreddit was diffuse, and bound by no strict order or cultural ideal, but like the Goons they shared a common social network. Over its first few months, Dreddit's founder, Fletcher Hammond, had built the original corporation into an entire Reddit alliance cheekily called "TEST Alliance Please Ignore," a callback to the most popular post ever on Reddit at the time "TEST post please ignore" which quickly exploded in popularity as thousands of people arrived to reply "don't tell me what to do."

With his labor complete Hammond stepped down and appointed a new CEO of TEST, his successor: "Dank Nugs." The reign of Dank Nugs would last only a few months before passing to "Montolio" who would define the organization's direction for years to come.

Shortly after Montolio took control, TEST Alliance membership surpassed 3000 players, and a new mascot was adopted. After testing the limits of CCP's patience with the first design, TEST eventually settled on a more respectable-looking version of its original concept: Middle-Management Dinosaur. TEST had found an identity that could unite a disparate social network, and there was no telling how large the alliance might grow as its leaders charted the nascent organization's future.

2011 EVE

While the Northern Coalition was blossoming into the largest power of 2010, it's no coincidence that the Drone Regions were a different story.

Drama had gripped the Russian coalition in the Drone Regions though nobody at the time in the Western community seemed to know what was going on. With the benefit of hindsight we now know that there was thick drama dividing the top Russian leaders, but at the time *EVE Online* knew Mactep, Death, and Silent Dodger (Red Alliance's leader) as best friends and pillars of the community. It would take some time to find out why the Russians were ostensibly allies and yet were rarely ever seen together. The only thing that seemed to be able to get them to work together was getting revenge against Atlas Alliance.

In the southwest, IT Alliance was in turmoil as well after the failure of SirMolle's MAX 2 campaign put the alliance into a state of chest-thumping and finger-pointing. Meanwhile in the deep south, Bobby Atlas had decided that this moment of chaos was the perfect opportunity for Atlas Alliance to step out from behind SirMolle's diminishing shadow. ●

Below: A ship glides through displays of celebratory fireworks during a seasonal event. Above Left: Rumor has it TEST's first logo submission was an MS Paint drawing that was rejected by CCP Games. Above Right: The logo that was eventually adopted, Middle-Management Dinosaur, was a symbol intended to reference TEST's desire to automate the alliance and eliminate the need for archaic middle-manager-types.

ATLAS SHRUGGED

"My name is Oofig VanDoogan and I've been sent from the future to
warn Atlas of the impending Pandemic Legion threat."
— Oofig VanDoogan, forum poster

Deep in the Drone Lands, the Russians were biding their time.

After Red Alliance lost Insmother to Atlas Alliance, the Russians cloistered themselves away in their Drone Region citadel, a vast place that most of the *EVE* community had written off as a waste of time, unworthy of the cost of the ammo it would take to conquer. Most believed the Russians had been allowed to survive there simply because it wasn't worth it to evict them and take what many saw as the Russians' junkyard. Between October 2009 and July 2010, most of *EVE Online* hardly thought about the Drone Regions.

But even without the natural advantage of technetium moons, the Russians had built a stockpile of raw minerals worth trillions in the quiet of the notoriously vacant Drone Regions. In their decrepit refuge—a great black sea dotted by inhuman drone hives built of equal parts hollowed asteroids and twisted metal scavenge—the Russians were hard at work developing a weapon system to regain what had been taken by the Western cur Bobby Atlas.

It might seem like hyperbole to those who don't understand the reality and permanence of virtual places, but the Russians were utterly fixated on reclaiming their home and getting revenge on those who had dared dock their ships in Russian stations, dared raise their flags over

Russian outposts. During most of 2010–while Goonswarm was collapsing, TEST was taking its first steps, and Atlas expanded through the South—the Russians were building. For ten months they lay silent, quietly building a super-capital armada to rewrite the history of New Eden, built from scrap and yet more powerful than anything that had previously been thought possible.

The age when Titan-class ships were a rare sight was long over. After the Great War their production had sky-rocketed, and the wealthiest groups in New Eden—including Atlas Alliance—beamed with pride as they looked upon indomitable fleets backed by a dozen or more of these behemoths. CCP Games watched with horror as the players found ways to build more than a hundred Titans across New Eden in just the first three years. What was once conceived as a rare crown jewel for the mightiest empires was now a common sight for all nullsec pilots. In addition to being common tools in combat, there was usually a Titan parked outside most alliance headquarters stations just to serve as a logistics transport ship. Fleet commanders would pilot them on alternate characters and use their warp tunneling ability which helped shuttle fleets around the region with ease. Their leaders thought their borders were impregnable, for what force could possibly contend with coalitions of alliances that could summon such overwhelming strength? Many believed the scales had tipped in New Eden, that those who held power now might never be unseated. But there were those who wanted to prove that the established order could still be overthrown.

The fleet the Russians were building was a tool to do exactly that. Within a year they would shock the *EVE* community when they fielded an unthinkable supercapital fleet, more than 150 strong. Half a dozen of them flown by Death himself, like a great multi-boxed chariot.

ATLAS OF INSMOTHER

There are two main ways that I've seen Bobby Atlas characterized by his contemporaries. The first and perhaps most generous casts him as an extremely meticulous leader. He micromanaged every aspect of Atlas Alliance. He kept the alliance organized. He kept the corporation CEOs happy and rich, and he decided the alliance's fleet strategies.

The other characterization conveys the same story, but is somewhat less complementary. This version of the Bobby Atlas story tells of a jealous leader who would not dole out power because he was hoarding it for himself. Bobby wanted to be powerful, and to lead a powerful coalition, and nowhere in that vision of power did he see a place for a bureaucracy that might help the coalition stay organized and functional without his involvement. Both stories convey the same message: everything went through Bobby.

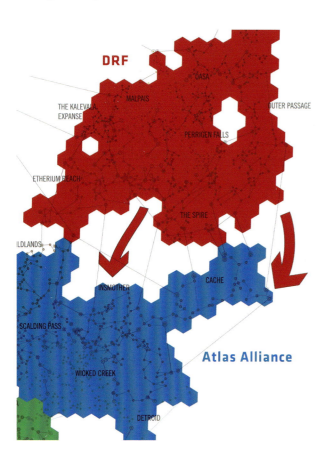

As his alliance grew to control half of the south, Bobby Atlas saw that his plan was succeeding. He was on his way to becoming one of the great leaders of New Eden, his alliance controlled huge swaths of territory, and the Russians—from his perspective—seemed bested. From the end of 2009 through the first half of 2010 an uneasy peace reigned through the south as both Atlas and the Russians contemplated their next moves. Bobby began to believe that it wouldn't do for *EVE's* greatest leader to remain a mere vassal of IT Alliance. He wanted to be the leader of his own coalition.

On July 20, 2010, Bobby Atlas sent a message to his allies in the Southern Coalition: Evil Thug (leader of Against ALL Authorities) and SirMolle.

"As of today my Alliance and my pets will be resetting the entire Southern Coalition," he wrote.

To "reset" is *EVE* shorthand for wiping the slate clean. All past political affiliations and alliances are dissolved.

"I thank all of my past friends, but now it is time to form my own coalition. Thank you BoB/IT Alliance, Against ALL Authorities [...] for helping us along the years," Bobby wrote. "With the recent and drastic reset hopefully nobody will be hurt by my decision but it has come to my attention that we have separate views on how we want to run our coalitions."

PANDEMIC LEGION

What Bobby Atlas didn't know is that just days prior, Death had quietly contacted the roving mercenary outfit Pandemic Legion that had been kicked out of its nullsec region by IT Alliance a few months earlier.

After a period of darkness and soul-searching, Pandemic Legion had found a new calling in mercenary work. Pandemic Legion was approached by Death and offered the contract of a lifetime to hit Atlas territory and generally make life hell for Bobby. As part of the contract, Death even gave Pandemic Legion several Titans to use in the effort.

Pandemic Legion was keen to use this as an excuse to stretch its legs and test itself against major powers again. Its figurehead fleet commander had returned to the game to reinvigorate the alliance. That leader, a player by the name of Shadoo—a veteran of the Coalition of the South that had fought Red Alliance in 2006—was helping to pioneer new fleet strategies as well, and he wanted to see how his new fleet concept would fare in a live war.

Pandemic Legion was an alliance of older players who had been in the game a long time. Atlas Alliance was much younger by contrast. In *EVE*, far more than other games, this can have extreme implications, because characters gain skill points in real-time throughout the months and years that they are subscribed to the game rather than by gaining experience killing rats and pirates.

Because Pandemic Legion's pilots had been playing the game for so long, they already had the skill points needed to fly just about any type of ship. Plus, they were quickly becoming reviled for their most insidious tactic: talent poaching. Every group I've ever interviewed who fought Pandemic Legion has told me stories of its best players being quietly groomed by Pandemic Legion talent scouts during the conflict. Any fleet commander or pilot who repeatedly distinguished themselves in battle against Pandemic Legion would often get golden offers to come play with the Legion and fly expensive, cutting edge fleet concepts with some of the most experienced veterans in the game.

By contrast, other alliances were made up of a hodgepodge of new players, old players, and everywhere in between. The main problem with this is that there was no way to know which players would be able to fly which ships when they joined a fleet. It wouldn't do to have every player in the fleet flying a ship that was meant to fight in a different way. Fleets needed to have coherent strategies to them in order to be coordinated with any kind of effectiveness. Trying to come up with a coherent strategy was often a logistical nightmare. Even if the alliance was wealthy enough to buy ships for players outright, there was no guarantee the pilots would be able to fly them. Fleet commanders and theorycrafters compensated by keeping their strategies loose and often downright informal. Up to this point in *EVE's* history the fleet strategies could be as simple as "everyone bring heavy armored ships with railguns," or slightly more coordinated with dictates like "bring sniping battleships" (long-range mid-class ships.)

Pandemic Legion realized that with a little bit of foresight it would be able to get hundreds of its members in extremely high-value, high-skill ships in a way that no other group of players was able to. Its theory was that they could get an entire fleet to fly the same ship with the exact same

Above: The avatar of the leader of Atlas Alliance, Bobby Atlas.

components, and unify its attacks and strategy in a way that would've been unthinkable a year prior. This simply was not the way *EVE* was normally played. Pandemic Legion's new "ArmorHAC" fleet concept, short for "Armored Heavy Assault Cruiser" featured hundreds of ultra-tanky Zealot-class hulls fitted with an expensive array of "Tech 2" components. In their final form they were also quick and good at fighting at close range. Those ships would be supplemented by repair and command ships that made them extremely difficult to kill, the ideal setup for fighting outnumbered. Ordinarily ships this expensive wouldn't be fielded because it would make reimbursing them too expensive for the alliance. By this point, however, many of the older veterans had grown wealthy enough that it wasn't as much of an issue.

On July 18, (just two days prior to Bobby Atlas' announcement that he would be resetting his allies) Pandemic Legion began its contract on Atlas and its pets. The first Atlas ally to be struck was called "Honorable Templum of Calcedonia." The result was utter havoc. Not only was the ArmorHAC fleet effective, it launched segments of the *EVE* community into a full-on panic. Not one ship from Pandemic Legion's new fleet was destroyed. It was so badly demoralizing that Honorable Templum of Calcedonia packed up and retreated immediately, and the more hysterical members of the *EVE* community screeched that Pandemic Legion had discovered some new hack or bug.

Worse still, the fact that none of the Pandemic Legion ships were destroyed was actually hugely important, because it meant that the losing side couldn't figure out what the fleet was made out of. Nobody could see the inner workings and reverse-engineer this new fleet concept.

Pandemic Legion had arrived in terrifying fashion, a fundamentally rebuilt alliance that was meant to be lean and fast and capable of striking anywhere at any time. But they had struck one of Atlas' pets, not Atlas itself. While they were Atlas' allies, they weren't exactly revered or respected. Most in Atlas assumed that the disaster was caused by the incompetence of those allies who had been put in that position specifically to be the buffer they were now serving as. Few in Atlas were too perturbed.

And so, two days after Pandemic Legion began its campaign, Bobby Atlas went ahead with his decision to spurn IT Alliance and Against ALL Authorities, and forge ahead with his own coalition.

ATLAS FALLS

There are episodes in the history of *EVE* with the flavor of a heroic epic; The Siege of C-J6MT, or the opening battle of the Great Northern War at P-FSQE, for example. But there are others which are the exact opposite.

The story of Atlas is important to the history of nullsec, but as I said at the beginning of this tale several chapters ago: it's anti-climactic. In fact, one possible reason why Atlas' history has largely been lost on the *EVE* community is because the story lacks a certain sense of climax. There's no memorable hook for people to place their hopes for the game upon, no great calamity that captured the community's attention. Just a punchline about fried chicken that will make Bobby Atlas into a mockery.

In July 2010, Atlas broke its coalition on the eve of a major invasion in a moment of admirable ambition and

Above: The avatar of Pandemic Legion fleet commander Grath Telkin.

regrettable hubris. In August 2010, that invasion began in earnest, and Atlas enacted a retreat-based defensive action.

"Red Alliance regained their strength and allied with other Russian groups and eventually Pandemic Legion, Red [Alliance,] and White Noise began pushing into Insmother. Atlas responded in kind and the battles in Insmother went back and forth for many days/weeks, especially the system C-J6MT. Pandemic Legion was employed. They came into the battle and at that point the battle tide turned from a stalemate to a full slaughter," wrote one Atlas Alliance pilot in retrospect.

Pandemic Legion was now being led by fleet commanders like Shadoo and Grath Telkin. While Shadoo was known to be a pretty mild-mannered guy in real life, Grath Telkin was anything but mild-mannered. A mountain of a man with a booming voice, wild grey facial hair, and long grey hair tied back into a ponytail, Grath was more like a folk character from a tall tale. Telkin gained infamy in the *EVE Online* community after getting into a tiff with another player on Reddit. When the situation threatened to escalate from a forum beef to a physical encounter at *EVE's* annual Fanfest gathering, Telkin used the opportunity to reveal he had a real life criminal past and would be more than willing to go down that road.

Grath Telkin remembered the first true battle of the war when Pandemic Legion finally managed to force an engagement.

"Their first major defeat (they had avoided contact) saw Dancul1001 and like 2 other Titans with a hand full of [supercapitals] augment a PL ArmorHAC fleet, a doctrine the game hadn't adjusted to, and saw that combined fleet fucking obliterate the much larger Atlas group. The space siege after that was Bobby trying to fight a war of attrition by retreating constantly and leaving the massive amount of space in the south as a buffer we had to burn through to get him."

"Atlas lost C-J and the war was squarely in Red Alliance/White Noise/Pandemic Legion's favor," wrote the Atlas pilot. "Atlas retreated to our main hub of 0-W778 in Detorid while fighting other battles around the regions."

When Red Alliance finally regained control of the station in C-J6MT it was renamed to "RA CJ6 REBIRTH," and eventually back to "RA Prime" as they believed it always should be.

By all indications, Bobby Atlas' strategy was to burn out the momentum of the Russian/Pandemic Legion forces by ceding the alliance's vast territories and leaving them as an undefended wasteland that the attackers still had to take weeks to conquer. This often has a way of sapping an attacker's morale by forcing them to engage in boring infrastructure clean-up operations. This tends to reduce an attacker's fleet force because its pilots stop logging in to participate in boring gameplay.

While that territory was being conquered, the diplomats in Atlas Alliance spent that time lobbying their old neighborly allies Against ALL Authorities to forget the time Bobby Atlas dissolved their coalition and come save them from the Russians and Pandemic Legion.

Manfred Sideous, who had replaced Evil Thug as the leader of Against ALL Authorities, announced that he would oblige and come to Atlas' aid. Even though Atlas had reset its political affiliations, Against ALL Authorities still had a remarkable amount of bad blood with Pandemic Legion, and Sideous felt forced to defend Atlas against unfair odds.

Against ALL Authorities sent fleets to aid its embattled ally, but when it arrived it found chaos. There was no clear indication of who was in charge of Atlas' forces, Bobby was AFK, and communication was non-existent.

There was a problem at the core of Atlas Alliance, and it was Bobby Atlas himself. As things began to get tough he simply wasn't logging into the game as much as he used to. He was becoming disillusioned with the entire game as the unstoppable Pandemic Legion and Russian fleets crushed his former homeland. After seeing the state of Atlas' leadership, Sideous called off the rescue and went home.

"At that point Atlas was dead on the inside, and it's impossible to save someone when they aren't willing to help themselves," said Against ALL Authorities' Manfred Sideous. "Most of Atlas's leadership were missing. The invasion took them by surprise."

By now Atlas Alliance's achilles heel had become blindingly obvious: it was Bobby.

"Bobby Atlas ran the alliance with a small group of directors and FCs," wrote that same Atlas member. "When Bobby was AFK, Atlas suffered."

Atlas Alliance had run headlong into a classic problem with the dictator model of governance: what happens if the dictator stops putting in the hours?

"They completely abandoned all their vassals at that point, and turtled up in O-W and Pandemic Legion and the [the Russian alliances] set up towers in-system and moved in for the kill," wrote Grath Telkin on Reddit in retrospect.

Though the rest of the *EVE* community had already moved on and considered Atlas Alliance a lost cause, within Atlas there was still a small spark of hope. Bobby Atlas came back.

After a significant amount of inactivity in the preceding weeks, Bobby was being seen once again in the alliance channels and he began hyping a great last stand as the Pandemic Legion/Russian force bore down upon the Atlas capital and formed a camp of artillery ships.

The next evening, Bobby Atlas rallied more than one thousand Atlas Alliance members—a massive turnout—to defend Atlas' home base as the final reinforcement timer elapsed and his capital station at O-W778 became vulnerable. This was Bobby's opportunity to turn back the tide, and put a stop to the destruction of all he'd sought to build as the leader of Atlas. The great flotilla of ships lumbered out from the capital to engage the invaders and make a statement that would define Atlas Alliance throughout the rest of *EVE* history.

The clock ticked down, the station became vulnerable. The Russian and Pandemic Legion force had swelled to about the same number of pilots but backed by dozens of Titans and several lethal ArmorHAC fleets, each 256 pilots strong. Bobby Atlas gazed upon the unstoppable armada that had come to conquer his home, and he gave the order.

"Stand down," he typed in fleet chat. "I'm going out for fried chicken."

BOBBY'S CHICKEN

"Morning came and Bobby had a massive Atlas turn out and faced a huge combined fleet of PL and DRF forces," wrote Grath Telkin in retrospect. "At the moment the timer came out, Bobby stood down and literally went out for fried chicken. Not even kidding."

There are moments in *EVE* that are so human they manage to somehow surpass the surreal. Reality, I've learned from *EVE,* is stranger than fiction, and the internet is stranger than reality. But to me it all makes a certain human kind of sense. What else was there for Bobby to do? Several nullsec leaders I spoke to for this book described the pain of what it's like to lose in *EVE Online.* How devastating it feels to build something for years and years only to have it ripped down—and your friendships ripped apart as you become more and more powerless to stop your worst enemies from exerting their will upon you and your people. How liberating it must have felt for Bobby to just let go, step away from the keyboard, and walk out the door to find some fried chicken.

"This whole war came to its climax at the battle for 0-W," wrote a different Atlas pilot. "Bobby logged in and was able to muster up about 1000 people with our enemies doing the same. Our [Titans and supercarriers] were logged in ready for an all-out brawl. Then Bobby just told us to stand down. (I was salty.) No fight was had and that pretty much sealed the fate of Atlas."

"0-W fell not long after and Atlas' market hub and financial might was crippled," wrote another Atlas pilot in a retrospective Reddit thread.

As the Russians pushed past Detorid and into Atlas' last vestige in Omist, the new leaders of Atlas who had tried to pick up the pieces in the immediate aftermath of Bobby's departure decided to surrender and ask for a treaty. Having been so thoroughly bested militarily, there was no doubt that their remaining territory was only Atlas' because there was so much of it that the Russians needed more time to conquer it all. Atlas was in an extraordinarily weak bargaining position. It's temporary leaders had no leverage, and the resurgent Russians took them for all they were worth at the bargaining table.

Who is Chribba?

A brief aside before you read the terms of the surrender, which may help you understand one oddity.

The terms mention both sides depositing "collateral with Chribba." Chribba is a famous player in the *EVE* universe who had made a reputation for himself as the only person who could truly be trusted in negotiations.

Warring parties or people conducting large and risky business deals who were wary of being scammed would often contact Chribba to serve as a broker who ensured that both sides complied with the terms of the deal. When expensive items need to be sold or transferred, Chribba is like the neutral dealer who shows up with the diamonds in a briefcase handcuffed to his wrist. A fascinating but necessary occupation for a virtual universe that is so often predicated on deceit and no-holds-barred domination.

"Earlier this morning a ceremony was held on the forward deck of Noobjuice's Titan. Present were such dignitaries as Campaign Commander Fintroll, Backup FC Shadoo, Morale Officer Jogyn, and Commissar Krutoj. Vice Admiral Ray Butts was also in attendance.

Bobby Atlas, feigning illness, failed to attend and Atlas Alliance was represented by Dastommy, who signed the following agreement:

Quote:

This war has ended today with the surrender of Atlas alliance to the Russians.

Both sides have deposited collateral with Chribba to ensure compliance [editor: 65 billion ISK and 115 billion ISK respectively,] and Atlas has agreed to the following conditions:

-Atlas will provide 2 Titans and 3 supercarriers to the Legion of xXDEATHXx/White Noise/Red Alliance coalition

-Atlas will drop the Atlas Alliance name

-The Russians and PL will set temp blue status to Atlas to allow them to evacuate their old space

-The remaining Capital Ship Assembly Arrays active in Omist will be allowed to complete their production

With the complete destruction of Atlas Alliance, Pandemic Legion has begun to de-mobilize from the region. [...] The somewhat poorly named VD Day (Victory in Detorid) has been marked with debaucherous celebration amongst the Legion and their Russian comrades, as Atlas forces stream outwards from their fallen capital [...] to their refuge in Omist.

To all former members of Atlas, the blind greed of your leadership led you to this point. We wish you a peaceful future as non-combatants of the Russian bloc. Farewell."

Phreeze, Pandemic Legion
August 29, 2010

The extraordinarily ostentatious deal stipulated that Atlas must disband its alliance, give the Russians a quarter of its surviving supercarriers and Titans, and sever ties with any allies they had left. In exchange, its leaders would be able to keep the rest of their Titan and supercarrier fleet, and some of their corporations would be allowed to stay in the deep south of Omist as vassals of the Russians.

The evening after the treaty was agreed upon, days after the failed last stand at O-W, Bobby Atlas finally returned, having caught wind of big events happening inside his alliance. He arrived amid an alliance in turmoil and mass evacuation, and from reading the alliance chat window—the logs of which were later leaked to the *EVE* public—he found out what was going on.

"A deal with Red Alliance, are you guys crazy," he wrote. "The minute Red Alliance get [our] supercapitals they're just going to continue the invasion with Legion of xXDEATHXx and friends."

Bobby Atlas was furious, but he didn't even know yet that his subordinates had agreed to shut down the entire alliance as part of the surrender.

"That is an actual term in the deal?" he asked, bemusedly, about the disbanding of Atlas. "Are you fuckin' joking? I will not sit by and watch a sellout to Red Alliance. I will sooner disband it myself. In seven days the alliance is disbanding."

It was already too late. Within days, Atlas Alliance lost 1500 of its 3000 members, and the rest left over the course of the next two weeks. With Atlas in a failure cascade, even its old friends Against ALL Authorities—infuriated that Atlas' directors gave Titans to the Russians for free just to save some of the directors' own Titans from being destroyed in battle—camped out Atlas' evacuation route back to empire space and hunted its retreat convoys.

"We called the deal for what it was—a treasonous act to all pilots in Atlas," wrote an Against ALL Authorities fleet commander named BlasterWorm. "We hoped Atlas leaders would reconsider but to no avail. Atlas was given 48 hour window to sort out their position. When this did not happen—we [cut diplomatic relations with] them. The vast majority of Atlas' allies were already making plans about the future without them. Death to traitors."

The Russians moved in and captured everything. From the factory in Tenerifis where SirMolle's first Titan "Darwin's Contraption" was destroyed, to the old wreck of Steve still floating in Esoteria. Hundreds of star systems including the Atlas capital at O-W778 were now occupied by Russian alliances like White Noise, Red.Overlord, Legion of xXDEATHXx, and Solar Fleet.

The lineage of Russian alliances now controlled almost all of the former RedSwarm Federation's territory, from the deep south of Omist up to the far northern reaches of Cobalt Edge. Most crucially of all, C-J6MT was back in Russian hands. The other leaders of nullsec were left to wonder, however: would the Russians be satisfied with their conquest of the south, or was this a community in expansion that would continue to take over *EVE*? The truth was that the Russian factions of this time were deeply divided, and had only come together to destroy Atlas which they perceived as a mutual threat.

Pandemic Legion was elated by the result of the campaign as its commanders now looked forward to a grand return to glory for the hellraising mercenaries backed by a surge of income, confidence, and Russian Titans.

The only other group in New Eden that could serve as a viable counterweight to the Russians and Pandemic Legion was the Northern Coalition, and it was too busy dealing with drama and raising up a generation of TEST newbies to pay much attention to the southern conflict.

Bobby Atlas never came back. Forum legend has it he's still out there somewhere. Searching for fried chicken. ●

TEST CHAPTER DO NOT PUBLISH

"One alliance stood out as a beacon of gibbering, unpredictable madness. Under pressure, OWN Alliance had become something completely novel, a unique drama vector in an aging and jaded galaxy."
— The Mittani, CEO, Goonswarm

Late one night, in October of 2010 as Tau Ceti Federation left the north and Goonswarm took command of Deklein, a player named Teredrum was fuming in the local chat channel of his alliance's home station in CU9-TO.

Teredrum was the leader of OWN Alliance, one of the former guests of Tau Ceti Federation who were allowed to live in the Deklein region under the new rule of Goonswarm. He had a reputation for an unpredictable temper, but tonight it was worse than usual.

He told his members to ignore the dozens of non-combat freighter ships that were orbiting his headquarters offering free ships and sympathetic words. The commander of the fleet of haulers—the famously non-threatening Badger-class, to be specific—said they had heard about the plight of OWN Alliance, and were here to offer what help the philanthropic pilots could to the downtrodden.

The pilots in the haulers were from the newly minted TEST Alliance Please Ignore who were flying in OWN's space spamming messages through the system-wide chat channels about wanting to help get OWN back on its feet. The cargo holds of the Badgers were stuffed to the brim with deconstructed Rifters—traditionally thought-of as the cheapest ship hull in *EVE*–and were offering them to OWN pilots without charge out of the goodness of their hearts.

Opposite: A fleet of Machariel battleships in the chaos of a large battle. Above: The logo and mascot of TEST Alliance, Middle-Management Dinosaur.

It was obvious from the outset that the gifts were tinged with more than a little sarcasm. One TEST member managed to fight through the giggling long enough to skillfully maneuver their Badger in front of the docking port of OWN's space station. When it was just outside the undock, the TEST pilot initiated the self-destruct sequence. An escape pod blasted back toward the fleet, and a single firework flower popped in front of the docking port before scavengers swiftly arrived to loot the wreck and found nothing but garbage-tier items. The pilot later said it was "to better distribute the supplies."

By now it was obvious that this "OWN Aid Flotilla" was an extremely elaborate prank. But it was a prank done to send a message to the ~1500 members of OWN: we have reason to believe Teredrum has stolen more than 100 billion ISK from your alliance.

REPLACEABLE SHIPS

Teredrum was furious about the stunt, and he thought he knew who was really behind it. Not Montolio, leader of TEST Alliance, but The Mittani, leader of Goonswarm. Montolio was in charge of TEST which had executed the stunt, but it was The Mittani who had invited the young TEST Alliance to live in the north to begin with. To Teredrum, TEST and Goons were two names for the same thing: trolls. Not "Elite PvPers."

There was some truth to that viewpoint, though Goonswarm and TEST were still entirely separate communities. While Goons had now been in *EVE* for many years and had since undergone genuine hardship, TEST was new to the game and fresh with enthusiasm. This had a way of fostering a certain joy that is often missing in a traumatized star cluster.

It was a joy that reminded a lot of people of the joie de vivre of the Goons many years earlier when they had first arrived in the game—before the Great War and the Karttooning of Delve—and in the eyes of many they were natural cousins. Trolling, griefing, immature, giggling cousins, but cousins nonetheless, united in their belief that trolling was a force for good in this universe.

TEST was among the most junior of members in the Northern Coalition, and as such it was given management of just two of the over 200 technetium moons controlled by the Northern Coalition. But even these two moons were a generous gift. Thanks largely to the Northern Coalition's manipulation of the technetium market, Goonswarm's gift provided a suitable income for the large up-and-coming alliance of Redditors.

With the proceeds, TEST Alliance had even instituted a model of a program that had recently been pioneered by larger and wealthier Northern Coalition alliances: a Ship Replacement Program (SRP).

In this time, having an SRP was something that alliances advertised in order to attract members. It was sort of like how every bank in the USA wants you to know it's a member of the Federal Deposit Insurance Corporation (FDIC). An SRP was a symbol of stability, safety, and of member-focused alliances. It was a way of showing that the structure of the alliance existed to facilitate fun experiences for the average pilot, not the enrichment of a few stakeholders at the top of the organization. Not every organization was so concerned with sending this message to its members.

"THE MADNESS OF OWN"

When TEST found out that OWN had six technetium moons yet virtually no ship replacement program for its pilots, a "threadnought" (a forum or Reddit thread that gains a great deal of momentum) started to snowball.

On both Reddit and the *EVE* community backchannel site "Kugutsumen.com" (a third-party website created by the infamous hacker who discovered the cheating CCP developer T20) commenters were aghast that the members of OWN were expected to risk their own ships while flying on operations for Teredrum. Many players worried that his pilots were getting a poor impression of the *EVE* community just because they accidentally got recruited by Teredrum, someone the wider community was increasingly seeing as "a delusional ruler of a hermit kingdom," as the Mittani had called him in a piece called "The Madness of OWN" published on November 22, 2010.

On Reddit, the threadnought gained momentum until eventually TEST Alliance broadly agreed on which punishment for OWN was the funniest: TEST would begin offering a Ship Replacement Plan...for OWN Alliance's pilots. For two days only, any OWN pilot whose ship was destroyed could send the kill mail (an exhaustive receipt of the encounter) to TEST's bookkeepers who would check the details and issue reimbursements as appropriate. It was a slap in the face that gently questioned OWN's right to exist as a modern alliance, and beckoned its pilots to jump ship and join TEST.

Both OWN and TEST were minor members of the Northern Coalition, but by now the drama was thick enough that the entire star cluster was eagerly watching what was happening between these "New Goons" (TEST) and a defensive, cornered Teredrum who loudly professed ignorance about why his alliance couldn't afford a Ship Replacement Program.

OWN was in the game because it liked to fight. It was a "PVP-only alliance" which meant that its members were expected to engage exclusively in fighting other players or training and preparing to fight other players. In return for sovereignty in the North, OWN was expected to muster dutifully when a "Call to Arms" was posted on the coalition calendar.

Teredrum took this ethos extremely seriously. His pilots were his sole grip on power, and their skill in combat was all that made the alliance relevant compared to any other group of pilots, so he resorted to uncommon means to keep them focused on their sole pursuit.

Teredrum and other moderators in his leadership cabal supposedly enforced strict communication rules on the alliance's players. Players were discouraged from fraternizing with each other on the forums. People who were caught discussing things unrelated to player-vs-player combat were either punished or outright silenced. Players who so much as asked questions about mining, for instance, are said to have been scolded and told to focus on PvP. Teredrum's enemies allege that players who made mistakes in live operations were often pulled

Opposite: The Badger-class industrial cargo-carrier in all its glory.

into private voice comms servers where they would be berated with raised voices. Players who questioned why leadership was silencing them were called spies who were fomenting dissent in an otherwise harmonious alliance. *EVE* is not always fun and games. Numerous examples exist of leaders who turned abusive once they were too powerful to be held to account.

When Tau Ceti Federation was strong it had been able to constrain the worst instincts of Teredrum; rumor has it that TCF once even forcefully destroyed a previous iteration of OWN Alliance to breakup what was seen as a malignant crew. But as Tau Ceti waned in power, Teredrum was able to remake the alliance more or less as it was before.

Though OWN was a bit-player even by the standard of regional politics, this small amount of authority quickly emboldened Teredrum without a healthy Tau Ceti to serve as a counter-balance.

Inside this small set of about twelve star systems, Teredrum was a king who could not be questioned because he controlled the technetium. So long as he controlled his six technetium moons, he was able to maintain that grip on power within his alliance.

That began to change when TEST started taking a closer look at that technetium money, wondering why OWN couldn't afford to buy its members ships.

AUDIT

TEST Alliance mobilized, and swarmed into OWN territory not with warships, but with that most deadly of *EVE* weaponry: spreadsheets. Every single starbase and moon mining operation was scouted out and documented. The bookkeepers of TEST Alliance figured out the exact amount of money that OWN should have been getting from its technetium moons and all its other sources of cash, and when it was all calculated and the math was made publicly available, it pointed to a scandalous conclusion: OWN Alliance should have had 100 billion more ISK than it was claiming it had. In 2010, that amount of money could have bought several Titans, which OWN clearly did not have. So the question rang throughout all of *EVE*:

Where's the money, Teredrum?

By now most of the *EVE* community was aware of what was going on in this tiny slice of space just north of the Cloud Ring. Though this had begun as a mere squabble between two allied player groups ("blue drama," something that is a daily concern for all alliances) by now it had attracted the attention of tens of thousands across *EVE*, and as the new leader of the Deklein Coalition (of which TEST and OWN were a part) The Mittani could no longer ignore the issue.

Teredrum insisted that this was all just a dark plot by The Mittani and his troll pet TEST to destroy the Northern Coalition from the inside. First, he said, Tau Ceti had *coincidentally* collapsed after the Goons' arrival. Next, they would destroy the military backbone of the north, OWN! And soon, The Mittani will reign over the entire Northern Coalition, and march against all of *EVE!* Ironically, though nobody believed him, Teredrum's vision of doom would not be far from the truth.

Teredrum even posted chat logs on the forums showing The Mittani apparently confessing his grand machinations to him in private, but they were visibly doctored and written with Teredrum's imperfect English. Teredrum wrote in excellent English for a non-native speaker, but as these things go, most fluent English readers would be able to discern that the writer of the logs was almost certainly not The Mittani, whose own language is not just fluent, but—as in his account below— often exactingly precise.

"I realized that things were going to go hellishly wrong when I first had a chat with OWN's leader, Teredrum, on TeamSpeak," The Mittani later wrote. "I've had personal conversations with a whole spectrum of *EVE* players over the years, and while some were obnoxious nerds, I had never encountered someone who left me feeling truly disturbed until I met Teredrum. Veering wildly between outright hostility and sneering obsequiousness (an emotional tone which I didn't even realize existed), Teredrum managed to use the phrase 'with all due respect' at least ten times in as many minutes, as if it were a verbal tic. The content of the conversation wasn't much better: Teredrum looked down on the other Deklein Coalition members with absolute contempt for not being 'pvp focused' while seemingly asserting OWN's independence, despite their 'guest' status."

The Mittani asked Teredrum to submit to an audit of his alliance's finances which could've been easily achieved by handing over his access code (called an API key) which would allow others to use out-of-game software to ask the *EVE* server what Teredrum had been up to.

There were two possible scenarios which the Deklein Coalition was worried about: either Teredrum and his leadership had used those funds to buy ships for themselves personally, or they had sold the ISK out-of-game and sent it off to the highest eBay bidder. Then again, maybe the information would exonerate him, and prove that TEST's calculation was wrong or that simple incompetent economics were to blame.

Without the API key, it was impossible to know for sure, and so Teredrum's repeated refusals began to look more and more suspicious. The Mittani later wrote that these refusals forced his hand. He couldn't just let Teredrum go now that he looked so damn guilty (whether or not the worst accusations were necessarily true,) and worse, he was asserting OWN's independence and refusing to cooperate with inspections. To let Teredrum off the hook would be to repeat the mistakes of Tau Ceti Federation, which had lost control of its guests.

For The Mittani, this was an opportunity to make a public example of someone. To show the *EVE* world that this was a more mature Goonswarm. This new organization stood for order and control, and The Mittani wanted to make clear that in his fledgling Deklein state he would no longer be tolerating dissenters or financial trickery.

So he did what he thought was just: He commanded OWN Alliance to leave Deklein, but gave them a chance to resettle elsewhere in the North. It wasn't that bad of a deal for OWN, except for the fact that this new territory was on the border with the Russians. If it was PvP Teredrum wanted, he'd find plenty of willing combatants at 4am in the Drone Regions.

```
"OWN alliance is being evicted from
Deklein for their leadership's repeated
refusals to submit API keys to explain
their breathtaking financial malfea-
sance. Vuk will be finding them a new
home on the eastern front.
     Over the past two months the lead-
ership of OWN have demonstrated a per-
sistent resistance to working with us in
the region when not outright conspiring
to try to poison us to the rest of our
allies.
     Despite weeks of patient restraint
on my part, my tolerance has reached
its end. OWN will be decamping to Venal,
or wherever; I don't give a fuck. They
have a week to evacuate their member-
ship's assets and must commence de-so-
ving their territory immediately. If
they resist or attempt to fight they
will be reset and purged with extreme
prejudice.
```

OWN had been, diplomatically speaking, owned. The TEST Aid Convoy was intended as a funny poke at a nearby ally who wasn't doing enough for its pilots, but it had snowballed into the defining moment of Fall 2010. This collision of egos was emblematic of this era of Northern Coalition politics. The Mittani and Goonswarm had come into ownership of one of the wealthiest territories in the game when Tau Ceti Federation left, and they weren't about to be pushed around by someone they viewed as a bit player. The Swarm was looking stronger and wealthier than it had since the Great War, and Montolio and Vile Rat were cultivating a close bond between what were now two of the largest alliances in the game. The Goons became veteran teachers and mentors to the flock of Redditors.

Above: A portrait of the avatar of Montolio, a major transitional leader of TEST Alliance Please Ignore.

Though largely still newbies, TEST's Aid Convoy stunt was essentially a galactic debut for the alliance and, critically, it was also a joke that put TEST on the good side of The Mittani. Though The Mittani says he did nothing to engineer this event, he was nevertheless likely very pleased that TEST had single-handedly manufactured a way for him to get rid of more members of the old Deklein Coalition in a way that left his hands clean politically while allowing him to look the part of the stable, measured ruler.

Tales of the TEST Aid Flotilla seeped out beyond the bounds of *EVE*, attracting even more players to try *EVE* and to fly for TEST. Behind-the-scenes Montolio was working day and night to forge the alliance into a proper nullsec power beyond its relationship with Goonswarm.

It seemed nothing could stop TEST's rise. As the story of TEST and OWN spread virally across the internet, the call rang out. Even those who had never before heard the names Montolio or TEST howled through tears of laughter:

"TEST ALLIANCE. BEST ALLIANCE." ●

THE SOLAR LEGION OF RED NOISE

"Six months ago, the commentariat was espousing the consensus that the Northern Coalition was almost embarrassingly invincible, that particular flavor of power which causes its enemies to beg for CCP to intervene on their behalf; now that same group heralds the Northern Coalition's imminent doom. What the hell happened?"
— The Mittani, CEO, Goonswarm

By the Fall of 2010, the memory of a time before the Great War was fading beyond recognition. Nearly two years since the war's conclusion, the details were starting to get a little fuzzy. For the players of the Northern Coalition at this time the principle story of the Great War was not about Band of Brothers. It was a story about how a malicious force had tested the might of *EVE's* true fortress in the north, twice, and even after supposedly conspiring with CCP Games to conquer New Eden, Band of Brothers had failed to breach the walls of Vuk Lau's "Fortress Tribute." The North saw itself as a battle-tested fortress, the stability of which was beyond question.

Counterintuitively, this dominance made the Northern Coalition into a relatively docile superpower. It had no ambition to continue to conquer territory unless that territory belonged to people who were once involved in Band of Brothers.

"The Northern Coalition's political stance was always uncompromisingly hostile towards the [Greater Band of Brothers Community] to such an extent that it often used that relationship to define other standings," reads the *EVE* wiki's entry on the Northern Coalition. "At the height of The Great War of 2008-2009 the Northern Coalition openly operated a 'With BoB or against BoB' politics."

As long as you had opposed Band of Brothers when it mattered, the Northern Coalition had no problem with you. It lived in what it believed was a sort of post-history era in which the great oppressor of *EVE* had been overthrown. Now the people of *EVE* could live in peace, free from threat, fighting only for fun. Best friends forever.

Opposite: Debris floats through Drone Region space in the border system LXQ2-T in Etherium Reach. Above: The logo of Northern Coalition member, RAZOR Alliance.

The Northern Coalition traditionally saw itself as an alliance of carebears who simply wanted to play *EVE* in collaboration with a peaceful community. Though its alliances directly controlled one-third of the conquerable star cluster, 200+ technetium moons, and likely more than 250 Capital Ship Assembly Arrays in constant production, there was a sense of benevolence about the Northern Coalition's largely European union. The arrival of the Goons and TEST complicated that reputation. The reason for its docile nature, more than one person has alleged, is because the Northern Coalition leadership was making huge sums of money selling their wealth on eBay.

RMT

In 2003, it was common to call your enemy a "dishonorable pirate," but in 2010 the common slur was "dirty RMTer." RMT is an acronym meaning "Real Money Trading," and there wasn't a single leader in EVE who wasn't accused of selling ships and ISK on Ebay for cash. It's also hard to sort the true accusations from the weaponized ones, because often players were just trying to drum up an RMT scandal against their rivals to prove a player had broken the EULA and should be banned.

Vuk Lau himself was wrapped up in a scandal after some incriminating chatlogs surfaced alleging that he was openly discussing the "good old days" when he could sell a billion ISK for 60 Euro while nowadays he could only get 20. He claimed it was all a joke taken out of context, and to this day he denies ever selling ISK. With that said, that price does line up with the going rate of ISK at the time. He categorically denied the accusations when I asked him about it.

"I was one of the biggest opponents of ISK selling," Vuk Lau told me. "I even broke a couple of real-life friendships because of that. That said, I understand why people branded me as an ISK seller and partially I was to be blamed. The only thing people saw at that time was the Northern Coalition holding 200+ technetium moons. Also we had a lot of enemies without a proper reason to hate the Northern Coalition so I was the perfect target."

It was scandals like this which were slowly turning the Northern Coalition into the new BoB; an ancient and unimaginably wealthy coalition held in power by perceived corruption. RMT accusations had a way of inspiring a unique ire from the community, and it was often the first attack on an opponent's reputation. Players from countries with lower-paying jobs were more frequently accused of the practice, even without proof, because the exchange rate difference can be massive. In 2010, 60 Euro went a lot further in Belgrade—where Vuk Lau lived—than it did in London, so the motivation was believed to be higher.

The true target of an RMT accusation is the alliance's membership. The message an accuser is trying to send is that an alliance leader is siphoning off funding for their own personal enrichment rather than using it to help the alliance as a whole. The hope is that some people will start to question their loyalty to their leaders.

There are others who think it's a good thing that some players were able to make large sums of cash selling their stuff outside of the game. They usually say it's because it turns those players into "event creators" who will plan long war campaigns and sacrifice their time to coordinate huge battles for their players to participate in so they can get their money. Usually that argument tends to come from people known to be the ones getting the money. CCP has never agreed, and has always maintained that real money trading is bad for the health of the in-game economy and harms average players by inflating the currency and diluting people's savings.

GUDFITES

As the Northern Coalition's power grew there was only one group in *EVE* that truly made them nervous: the Russians. Many of the groups in the eastern half of the Northern Coalition (as opposed to the Goons, TEST, et al in the West) also happened to be Eastern European, and there was often nationalistic friction along the eastern

boundary as Serbs, Poles, Slovaks, Ukrainians, Estonians, and Russians slung insults and threats back and forth across the border dividing the regions Geminate (Northern Coalition) and Kalevala Expanse (Legion of xXDEATHXx.)

The Northern Coalition didn't have complete information on what the Russians—particularly Solar Fleet—had been doing all of this time, but after seeing what the Russian community and Pandemic Legion had done to Atlas Alliance, the Coalition became convinced that the true purpose of the Russian supercapital armada was an eventual invasion of the North.

The Northern Coalition was rarely involved in territorial warfare, so Vuk Lau opted instead for a preemptive strike that would hit Solar Fleet's Capital Ship Assembly Arrays in an attempt to slow down its supercapital production. Since this wasn't a full-on attack, an inexperienced

Above: A portrait of the avatar of Vuk Lau, leader of the Northern Coalition in 2010. Opposite: A Northern Coalition propaganda image created by Rick Pjanja featuring Joseph Stalin. Soviet imagery was a favorite tool of Northern Coalition propagandists.

fleet commander named Cobra2K was put in charge of the attack.

One Northern Coalition member spoke out against the attack, however. The Mittani and Goonswarm had grown friendly with many Russian figures during the Great War against BoB. The Mittani voiced a hesitation to ever confront the Russians or get involved in their internal affairs.

"I said, 'have fun with that,'" said The Mittani in an interview. "We don't fight Russians because that's dumb."

The Mittani believed it was unwise to risk provoking the Russian community in the delicate political sphere of nullsec. The main Russian alliances were divided and squabbling right now though nobody yet knew why but The Mittani had worked with many Russian figures and alliances during the Great War and had seen what they were capable of when properly motivated. The Mittani said that provoking the Russians was "just not wise." He advocated for containing them by keeping them fighting amongst themselves rather than united against an external threat to their community.

But Vuk Lau would not be moved, and the pre-emptive strike proceeded. With thousands of pilots rallying to his banner to slow the Russian threat, Cobra2K lead the charge into the East to take Solar Fleet down a peg.

But as he planned the campaign, Cobra2K realized that in order for these attacks to be viable, he needed to capture a system on the border called LXQ2-T in order to access Solar Fleet's territory. The system was owned by Legion of xXDEATHXx. He messaged Death and told him that this was not a serious attack, but rather a necessary staging point for creating fights against Solar Fleet's structures. A mostly good-natured shoving match between regional rivals (gudfites, as *EVE* players sometimes call it.) Death didn't hear it that way.

"Northern Coalition came to us and said, 'We want to destroy Solar Fleet, can you promise to stay out of it?'" said Death in an interview. "We said no. Because once you kill him why not go after me? It would have been a stupid idea to let him get killed. So I said no."

For days before October 30, Cobra2K was hyping his major offensive on LXQ2-T. This was an attack on a major strategic chokepoint into the Drone Regions, and he wanted as many people there for it as possible. Instead of standing aside, Death put out a Call to Arms of his own, spreading the message to as many pilots in the disunited Russian community as possible.

The rumblings of the impending attack were so big that CCP Games even heard the commotion, and their service technicians allocated extra processing power and server bandwidth to prop up what they expected would be a major battle. The wider gaming world was fascinated by the enormous battles in *EVE*, but in the recent past CCP had been embarrassed when it attracted a great deal of attention to a certain battle in Y-2ANO only to have the whole community watching when the server crashed.

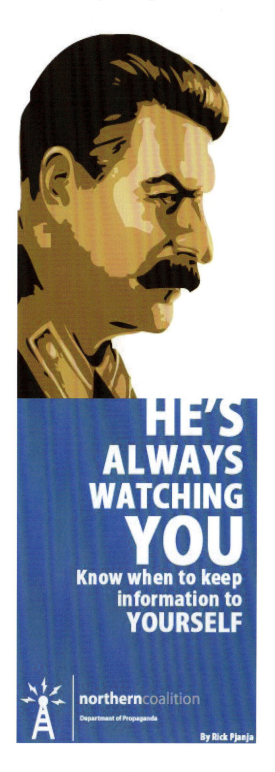

HE'S ALWAYS WATCHING YOU

Know when to keep information to YOURSELF

northerncoalition
Department of Propaganda

By Rick Pjanja

This time, they'd be monitoring the stability of the servers live and hopefully making fixes on-the-fly.

For ten hours prior to the iHub (the center of a star system's infrastructure, and the main focus of an attack) coming out of reinforced mode, Cobra2K piled as many ships into the system as he could. The day wore on, and the Northern Coalition's numbers continued to swell as the hours passed, eventually reaching a peak of around 2000 ships. The blob was finally revealing its full form.

The Russians arrived in LXQ2-T late, their own fleet numbering 982. It was a historic gathering, and yet the lack of unity between the Russian alliances placed them at a vast numerical disadvantage. Being massively outnumbered was the least of their problems, however. The Russians were facing a now classic *EVE Online* conundrum: will the server crash if we try to jump the fleet into an already crowded system? It was a steep gamble which had seen numerous fleets destroyed throughout *EVE* history, but the combined Russian fleet was too bullheaded to gather all these people together just to cede the system.

The Russians jumped their fleet into LXQ2-T, and miraculously, the game didn't immediately crash. Lag was crushing, but the server was operating and they could see what was going on—far more than many fleet commanders throughout *EVE's* history have had to work with.

Outnumbered 2:1, the Russians still felt they had a strong chance to defend this iHub. The Northern Coalition were supposed to be the "carebear" alliance with little experience in player-versus-player combat. The Russians, on the other hand, saw themselves as *EVE's* most fire-proved force, born in the crucible of the Siege of C-J6MT.

The ships finally clashed late in the day on the night before Halloween, and as the Russians refused to cede LXQ2-T to a much larger force, live reports came in through EVENews24.com.

RECORD BROKEN: 3100 + 600 IN LOCAL
"A new record has been set in New Eden. 3100 pilots at the same time in one system with 600 more trying to log in. Accurate numbers are hard to get, but this should be verified by CCP. LXQ2-T (Etherium Reach) brought Northern Coalition against Drone Russian Forces in the biggest clash ever in *EVE* history. The result is unknown atm, as lag is very heavy and no one can do anything about it. The servers are still going on, but it's a powerpoint presentation at the moment."
— Czech Lion, EVENews24.com

Numerous commenters on that article offered their own experiences as pilots inside the fight. Most of them just wanted to complain that they weren't able to log in, or that it was taking 50(!) minutes for their ship to get a lock on an enemy target.

The ships clashed late into the night on October 30, and over the final few hours before server downtime came around, more than 1200 ships were destroyed in a frantic melee. The losses were disproportionately Russian.

As the scheduled daily server shutdown time came around and the battle continued to rage, both fleets prepared to be simultaneously logged out. The powerpoint presentation came to a halt, and for 30 minutes both sides reorganized themselves for the next phase of the battle when the server came back online.

Thirty minutes later, pilots from both fleets were slowly getting back online in staggered numbers, breaking their formations. The battle continued with neither side willing to cede the field.

LXQ—OCTOBER 31—KILLING GOES ON AND ON
"LXQ system on October 31. Over the whole day, continuous fighting is happening. Many pilots got caught after logging back in in the aftermath of yesterday's monster battle. Several fleets have been chasing each other in and around the system. One staging tower down as well. Kills/losses are equal for both sides. There should be 74 Drone Russian [Federation] (DRF) supercarriers and 24 DRF Titans logged in system."
— John45, EVENews24.com

However, as the fighting continued, the Northern Coalition gained a grip on LXQ2-T and forced the Russians to log out their supercapital fleet.

More than 1200 shipwrecks littered LXQ2-T, and the Northern Coalition celebrated a historic victory. Its leaders bathed themselves in glory as yet more evidence compiled of their indomitability and destiny. Perhaps, they thought, the myth of Russian power truly was just a myth. Perhaps they were so weak that this playful campaign into the Drone Regions might incidentally cause the Russians to fall apart once and for all.

In truth, this victory was a double-edged sword. On the plus side, it demonstrated that the Northern Coalition blob reigned supreme in nullsec fleet combat. However, it also incidentally demonstrated—from the Russian perspective—a huge level of commitment from the Northern Coalition.

What Vuk Lau and Cobra2K hadn't taken into account was that the Russians had a completely different perspective on this situation. From their point-of-view, a huge number of Northern Coalition pilots were flooding over their border in military hardware. Viewed from the Eastern half of the map, the Russians saw an all-out invasion that was destroying their rental fiefdoms and damaging multiple members of their coalition; an enormous threat backed by the largest and most powerful coalition in the history of *EVE*. Vuk Lau's repeated assurances that this invasion was nothing to worry about looked to Death like deception and ambition.

Because they entered into this fight with relatively innocent intentions, the Northern Coalition had no understanding of how existential this threat appeared to the Russians. It's even possible that some of those Russian leaders understood this was not a full-on Northern Coalition invasion, but saw some personal benefit to using this crisis as an excuse to bridge old divides in the community. In particular, multiple sources suggest that PsixoZZ Kahi, leader of White Noise, had an axe to grind with the Northern Coalition, and wanted to use this to spark a full-on Russian invasion. Some would later allege that he was an agent of a rival ISK seller who was hired to take down Vuk's technetium monopoly.

PsixoZZ moved White Noise to the front line and became infamous for diving his ship into the middle of Northern Coalition fleets then typing things like "guys come save me" in Russian, just to trigger battles when others were too tentative.

As evidence mounted of the Northern Coalition threat Red Alliance joined the front as well in a show of solidarity with those who had sheltered them during the Atlas Alliance attacks on Insmother a year earlier. Even the reclusive Solar Fleet pledged assistance, but its own members would later say it was mostly symbolic. "The fifth wheel of a cart," they called Solar Fleet.

Above: A portrait of the avatar of Imperian. lead fleet commander for the Northern Coalition.

The Russian alliances retreated after the battle for LXQ2-T. The four leaders of the principal Russian factions realized that the Northern Coalition threat was far more serious than they had believed. Furthermore, they were disgusted by what they thought were craven tactics from the Northern Coalition. The Russians saw their 2000-person fleet, and felt like the Northern Coalition was blobbing the server to death before the Russians even arrived so that they couldn't have a true fight in which both sides' fleets were working at full function.

"The thing about the Russians is you don't want to interfere in Russian politics, because when you do, the Russians get together and take it personally," said DaBigRedBoat of Goonswarm. "They may fight each other all year long and then that one day when someone [insults them] they get together and they go, 'hmmm... fuck you.' That's what happens."

"We had three main alliances that were working as part of the same machine," said Death about Legion of xXDEATHXx, White Noise, and Red Alliance. "We took away all of our ambitions. We took away all of our differences. At the first losses, we sat down and tried to figure out what are the weak points, why are we losing? So, at this meeting, three of us agree that I'm going to be the political leader. PsixoZZ is going to be the general, the pusher, the field general, the guy on the field who inspires people. And the Dodger, he was more like a backup guy for bringing supplies, managing small stuff and shit like that. That's how we agreed. I was the strategic guy. Making decisions about how to do it and when to do it."

For reasons I didn't understand during our interview, Death didn't mention the "fifth wheel of the cart": Mactep and Solar Fleet.

Though the Russian leaders had been separated by messy internal politics for the last five years, they had now found a true common cause again. Here was the existential threat to Russianness necessary to unite their clans.

Their leadership—Death, Mactep, PsixoZZ of White Noise, and Silent Dodger of Red Alliance—begrudgingly joined the same Jabber channel again, and began making joint plans amongst themselves to restore stability to the Russian renter farms and subdue the carebear threat. They called themselves the "Solar Legion of Red Noise," but the rest of EVE would come to know them as the DRF: the Drone Region Federation. Or "Drone Russian Federation." Nobody could ever agree.

FATIGUE

The newly united Russians believed strongly in their ability to ultimately win this war in the long run, but they needed time. Their pilots had just been through the war with Atlas Alliance, and while that was a great morale boost, it also had an effect on their battle readiness because many of their best pilots were fatigued by the drudgery. The later months of the campaign had turned into a long series of cleanup operations to sweep the area of old Atlas stuff.

This "fatigue" is an interesting word in EVE parlance. It doesn't exactly mean the same thing as battle fatigue would ordinarily imply. In practice it's more akin to what people refer to as "burnout" in the tech and video game industries. Jobs that require lots of computer usage often cause an analogous symptom. The person often loses any sense of joy in what they're doing, and taking any virtual action feels like walking with concrete shoes. Irritability skyrockets and patience plummets. The average Russian member needed a break in order to reach their full potential. Usually when players in EVE feel this way, the solution is simply to stop playing the game for a little while. They use their spare time to catch up on work/school or be with their families instead of using it to show up for fleet operations.

The nascent Drone Region Federation needed time to get its pilots organized and its war fleet to the northern front, but time was exactly what it didn't have. Thousands of Northern Coalition pilots were already bearing down on its strategic crossroad systems. The Russians needed help.

SHADOO

It was at this moment when Pandemic Legion—the mercenaries who had served the Russians so ably in the war against Atlas—approached both Vuk Lau and Death with an offer. They'd seen the reports of battles in this area, and they wanted in on the gameplay content. The offer was simple: "we'll fight for the highest bidder." The most effective mercenary fleet in New Eden was now a rogue piece on the chess board that either coalition could purchase.

"When they offered to 'assist' us, at that particular moment I disliked the team leading Pandemic Legion as they were all but honorable, and spaceship honor was the only thing important to me in EVE," said Vuk Lau. "They were trying to play both sides bargaining for extra ISK."

The Russians, on the other hand, were flush with cash from their massive renter empire and had no qualms about making an obvious choice. Vuk Lau saw extortion; Death saw a lifeline. He had seen first-hand the devastating effectiveness of Pandemic Legion's new close-range Heavy Assault Cruiser fleet strategy, the "ArmorHAC," and had even given Pandemic Legion Titans to use against Atlas Alliance. The Russians didn't want to be on the wrong end of their own Titans. They pooled their funds and signed a massive 600 billion ISK deal with Pandemic Legion. In exchange, PL agreed to lead their war effort.

The Drone Region Federation, wary of trying to share strategic command between the four of them, gave Pandemic Legion's Shadoo control over the battleplan. Shadoo devised a plan to re-establish a defensive front

line in the Drone Regions while driving the tip of the spear deep into the Northern Coalition's wallet. All of this would be combined with a political campaign spearheaded by Death designed to split the Northern Coalition in two: the Deklein Coalition led by TEST and Goonswarm, and the western allies led by Morsus Mihi and Razor.

The first stage of the campaign was a two-pronged attack. While the Russians prepared to take back LXQ2-T and regain control of their lost system, Pandemic Legion went behind enemy lines into the Venal region and struck where it would hurt the Northern Coalition the most: its technetium moons.

Venal is a region of space that can't be officially conquered by players, but infrastructure can still be built and attacked and resource collection can be shut down. Mining towers would shut down and cease operating if they were attacked by a fleet, then come back online 36 hours later and resume collecting the "moon goo" that formed the backbone of the Northern Coalition's finances. Pandemic Legion methodically shut down every technetium-producing moon every 36 hours and established a firm military hold on the area, stopping the flow of technetium to the Northern Coalition members.

As the defeats piled up, the Northern Coalition realized that Pandemic Legion's ArmorHAC fleet concept couldn't be beaten. It could only be matched, and the Northern Coalition began trying to imitate its success. But Pandemic Legion's fleet concept had a secret weapon that nobody could imitate, because nobody knew about it: an extensive spy support network.

Spy networks were already commonplace in this era of *EVE*, but they were mostly used for intelligence collection and keeping tabs on enemies. Occasionally, an agent would see a particularly glaring opportunity to turn off some defenses or steal a fleet of ships, but for the most part their job was to relay information they'd read on the forums or seen in alliance chat.

Pandemic Legion's spies took things one step further. They infiltrated their enemies' organizations at the low-level, and would dutifully show up for fleet operations. These spies would simply listen to the enemy fleet commander's instructions and do as they were told. But simultaneously, they'd be in another chat window with Shadoo. As the Northern Coalition fleet commander barked out orders to focus the attack on certain Pandemic Legion ships, the spy would relay that information to Shadoo who gave the command to his support ships to start repairing that ship, often before the Northern Coalition fleet even began firing. When the Northern Coalition tried to ape Pandemic Legion's ArmorHAC concept it was frustrated to find it wasn't working nearly as well, because the coalition lacked the invasive spy network required to make it work at full efficiency.

To make matters worse, Pandemic Legion quickly saw that the ArmorHAC was finally being copied so it developed a new fleet concept designed specifically to dismantle its own ArmorHAC. It was essentially the same idea, but this time PL would fit its ships for long-range engagements. While the short-range ArmorHAC fleet chased after them, the all-new "Hellcat" fleet kept them at sub-optimal range. As the Northern Coalition ArmorHAC fire continuously missed, the Pandemic Legion Hellcat fleet chipped away at them from perfect safety. The EVE community shuddered at the sheer domination on display in some of these battles.

The Northern Coalition was on its back foot trying to deal with these attacks behind its lines that were disrupting its moon income. The Pandemic Legion Hellcat fleets were unbeatable unless the Northern Coalition managed to summon overwhelming reinforcements (such as one battle in which their forces were reportedly supplemented by nearly 600 Goons to force off a Pandemic Legion Hellcat fleet of 100.) The Northern Coalition was hemorrhaging funds, and the battles were nowhere near cost (or time) effective.

The wider EVE community jeered at the Northern Coalition and said their "blob" was a craven strategy of population over skill. The Northern Coalition pilots usually took it in stride, and would snipe back something like, "the blob is coming for you next," and often signed with the ever-present alliance motto, "Best friends forever!" Suddenly, however, it seemed that quality was triumphing over quantity.

Occasionally, by sheer numbers, the Northern Coalition was able to push Pandemic Legion's sub-capital fleets off the field. All this meant, however, was that it was time for PL's supercapitals to arrive. The pulsing mass of Titans looked like a thing occult. Titans are intended to be the biggest ship in EVE, but a single Titan looks almost silly next to a fleet of them. With minimal ability to maneuver delicately, the fleet of dozens of ships mashed together, their 3D models clipping through all the others and forming a great spiky ball of ships slowly dissociating, occasionally causing one or two to glitch and bounce away into space. Pandemic Legion had a blob of its own, growing stronger all the time, and it knew how to use it better than any other alliance in EVE.

Throughout November and December both sides jockeyed for position on the borders of their former territories. The battlefield in the Drone Regions looked mostly the same, but notable success had been achieved by Pandemic Legion, which now claimed the moon-rich region of Venal as its own.

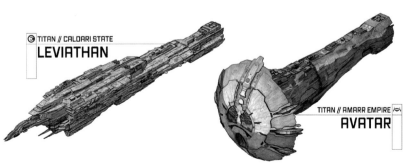

TITAN // CALDARI STATE
LEVIATHAN

TITAN // AMARR EMPIRE
AVATAR

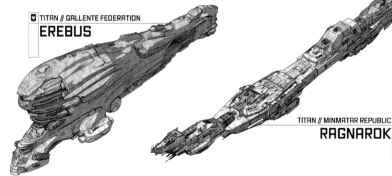

TITAN // GALLENTE FEDERATION
EREBUS

TITAN // MINMATAR REPUBLIC
RAGNAROK

Opposite: The four different classes of Titan all have somewhat different capabilities, but all have the use of the Doomsday main weapon. **Above:** A mass of Titans glitches together as a fleet masses for an attack.

"As of 23 December all Venal tech moons are owned by Pandemic Legion," a spokesperson wrote in a publicized announcement. "70 tech moons producing 90 billion isk a week. All alliances but the original Goonfleet will be hunted down and killed if entering Pandemic Legion's region. For renter agreements and buying the tech moons contact Pandemic Legion."

As 2010 came to a dramatic conclusion, the Russians were readying for the true assault. It would begin on December 29, 2010, when the Drone Region Federation unleashed its long-awaited supercapital invasion of Northern Coalition-held Geminate.

ATTRITION

The Drone Region Federation wanted to start off its renewed invasion with a bang, but early attempts were frustrated by failure.

On the first day, a fleet of 400 members pushed forward into Geminate and managed to trap a Northern Coalition defense fleet consisting of members from its border with the Russians. The alliance, called Rebellion Alliance, was non-coincidentally also largely Russian and had been placed there to guard the Northern Coalition in the Russian prime time zone. Once trapped, the Rebellion fleet put out a call-to-arms to the larger Northern Coalition community and two more fleets were coordinated which pincered the DRF fleet into the system and destroyed it.

On the very next day, in nearby O2O-2X, a DRF trap again went horribly wrong. Pandemic Legion was tracking a Northern Coalition fleet moving through the region, and wanted to ensnare and destroy it. The Pandemic Legion Hellcat fleet waited patiently near a Titan, intending to use it to jump the fleet right on top of the supposedly clueless NC fleet.

But the Northern Coalition fleet commander, by the name of Yaay, knew all about the trap, and called up additional capital and supercapital forces to be ready nearby. Then Yaay ordered the fleet directly into the Pandemic Legion trap. Pandemic Legion's Hellcat fleet jumped to optimal range and began trying to destroy the NC fleet commander. Yaay triggered the trap and brought down a full complement of dreadnoughts and Titans. But Pandemic Legion wouldn't blink. Shadoo saw Yaay's raise, and responded by fielding his own supercapital force.

Yaay directed his Titans to begin targeting known Pandemic Legion fleet commanders and logistics ships that provide buffs to the entire fleet. Shadoo responded by focusing Pandemic Legion's fire on one Northern Coalition Titan in particular, flown by Vuk Lau himself.

Vuk Lau's Titan was destroyed first, and another Northern Coalition Titan went down next, but reinforcements tipped the battle out of Pandemic Legion's control. In exchange for their two lost Titans, the Northern Coalition took down six Pandemic Legion Titans, an unprecedented sum even for a coalition (let alone a single alliance.) By the end of the day, Pandemic Legion was down 600 billion ISK, the entire amount of their contract with the Drone Region Federation.

In one grand maneuver, the Northern Coalition leveled the playing field against Pandemic Legion after being pushed around by its mercenary fleets for months. The legend of Northern Coalition dominance was still growing.

Throughout the first month of 2011, Pandemic Legion was able to maintain supremacy with its Hellcat fleets, and yet continued bleeding Titans largely due to innocent mistakes and clever Northern Coalition maneuvers.

In the Drone Regions, the Russians recaptured the last of their lost territory, and were able to join forces with Pandemic Legion on a campaign of retribution. PL wanted revenge for O2O-2X. The Russians were united. Both wanted to ensure the NC could never threaten them again. The campaign to safeguard the Drone Regions was now openly advertised as a campaign to destroy the Northern Coalition forever. ●

Above: Nyx and Aeon-class supercarriers near a huge player-owned station. Below: The scale of battle was constantly growing. Seen here, hundreds of ships gather to besiege an Infrastructure Hub, the center of control over a star system.

REBELLION

"In the spring of 2011 many of the Northern Coalition's long-time enemies joined forces with the Drone Region [Federation] superpower, and hired supercapital power Pandemic Legion to form a supercapital fleet massively surpassing anything the galaxy had witnessed before. Many predicted a repeat of former invasions, but this time the NC was spent early. The Fortress was finally broken."

— Morsus Mihi History Wiki

The course of *EVE Online*'s history—the ebb and flow of fleets and communities—has often been decided by just a handful of people. One of those people was Daroh, leader of Rebellion Alliance, a neutral group situated on the border between the DRF and the Northern Coalition as a sort of buffer between the two superpowers.

Though Daroh himself was not considered to be in a position of power, he had specific powers which were critical in their moment. Daroh originally brought the alliance together to have fun and find big fights and in order to do that he had to make alliances on both sides of the divide. However, the Northern Coalition was now asking more and more of its ally to defend the Russian time zone, and Daroh decided that facing down an armada of Russian ships was not what his alliance had originally come together to do. Daroh said that he felt "enslaved" to the coalition and the need to constantly be on-call whenever they were attacked. Daroh chose to disband his alliance in the face of a massive supercapital invasion rather than serve as the RUS timezone "meatshield" for the Northern Coalition.

"It seems that Daroh of Rebellion Alliance (-R-) has decided to disband his alliance this morning, leaving six stations in Geminate vulnerable," reads an article on EVENews24 dated February 13, 2011. "What makes it worse is that the Drone Russian Federation seems to be invading [...] starting today. The NC is currently scrambling a defense so we should once again see an escalation of hostilities in Geminate."

Opposite: Inside the fighter bay of a Minmatar Hel-class supercarrier.

Furious Northern Coalition pilots hurled accusations of bribery, saying that once again the infamous Russian EBay ISK Kingpins had bought off a fellow Russian with ill-gotten gains. Daroh himself posted a lengthy reply to the drama on the Russian forums, denying the accusations and giving his perspective. He wrote of how his alliance had come together to find a way to have fun in this strange universe, and that entanglement in complex alliances slowly killed off any sense of freedom that he and his pilots had. What follows is a translation of part of his 2000+ word goodbye letter to the community:

"Unfortunately, relations in the Rebellion Alliance Council have reached the so-called point of no return. I am not going to advertise the reasons for this (I don't want to pour shit, although I can imagine how much shit will pour on me now), but I can only say that disagreements arose over which I do not see my corporation's continued presence in this alliance.

The Rebellion Alliance was created by me and it will end its existence with my exit from the alliance. You can say that I had no right. That I threw people away. That I was scum and ruined everything that others built. But the members of this alliance should think about what they have, what they built, and how it affects their fun before blaming me for taking something away from them.

If the fish has already rotted almost to the tail, then it is better that

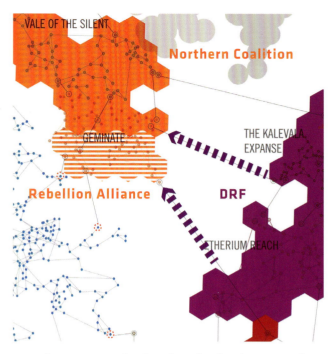

There was scarcely any time to debate the true cause behind what had motivated Daroh. Within hours of the disbanding of Rebellion Alliance more than 100 super-capital ships were seen slowly moving forward from DRF territory into one of Rebellion's abandoned systems in Geminate.

Fresh with enthusiasm inspired by the "traitorous" Daroh the Northern Coalition sent out an urgent Call-to-Arms. Legendary Fleet Commander Imperian would lead thousands of pilots into battle to retake Daroh's abdicated systems.

TITANS CAN'T DOOMSDAY IN LOWSEC

Imperian's plan was to allow the DRF to capture Rebellion's systems but set a trap for the DRF fleet on its way back home. However, as previously noted, the Russians and Pandemic Legion by now had an extensive spy network within the Northern Coalition fleet, and word reached Shadoo (Pandemic Legion) and Death (Legion of xXDEATHXx) about what Imperian (Northern Coalition) planned to do.

The DRF quietly assembled a Titan fleet in secret, and instructed them to log out of the game nearby, and wait for Imperian to bite the lure. As the DRF bait fleet began heading back to the Drone Regions after securing the last of Rebellion's systems the fleet jumped into the low-security system "Uemon." Imperian triggered his trap and jumped in the entire Northern Coalition fleet. Seeing a huge tactical advantage, Imperian committed 24 Titans to attack the DRF fleet.

As the Titans arrived, Imperian gave the order for three Doomsday weapons to target the Pandemic Legion fleet commander Shadoo. But instead of enormous beams of laser light converging on Shadoo's flagship, nothing happened. The fleet had completed its jump and the small subcapital fleets began to engage each other. Drones poured

out of supercarrier loading bays by the dozens, and yet Imperian's Titan fleet remained motionless. The trap had gone off without a hitch, and the NC fleet towered over the vulnerable DRF force. There was just one problem Imperian hadn't accounted for: Uemon isn't technically in nullsec. It's a border system in "low-sec," where Titans aren't allowed to use their Doomsday weapons. For most Titan pilots there was never a reason to bother flying their prized ship into low-sec so the issue rarely came up. Now, Imperian had brought not just his Titan to low-sec, but every available Titan in the Northern Coalition. Imperian realized his mistake, but it was too late. Russian Heavy Interdictor ships (warp scramblers) charged into the fray, and began attempting to scramble Imperian's most valuable Titans.

Cynosural fields ripped open in the center of his fleet, and out poured the Drone Region Federation counter-drop force with entire fleets of supercarriers and Hellcats that were far more capable of operating successfully in the Doomsday-less low-sec battlefield. Roughly a third of the Northern Coalition's supercapital fleet was stuck here in low-security space, as the DRF and Pandemic Legion seized their first good opportunity to break this war wide open.

The thousands of ships on both sides clashed together and became a chaotic mess. Viewed from afar it looked like a diffuse spherical mass of ships—most of them too small to see—slamming into each other. Amid the teeming mass of warships, every once in a while a civilian would wander through the chaotic scramble; not a miner or an innocent bystander, but a curious *EVE* player who heard about the fight in low-security space and wanted to see it first hand.

Above: Main Northern Coalition fleet command team (from left to right) Sala Cameron, Yaay, and Imperian.
Opposite: Daroh disbands Rebellion Alliance and creates an opportunity for the DRF.

In the list of casualties in these battles there are always a handful of the worst ships in the game, the ones you automatically get when you open an *EVE Online* account. These players create new characters just to fly disposable ships into the fracas, and get their names in the data of the record books. In the post-battle statistics detailing the destruction of Titans and supercarriers, there are also small newbie frigates flown by joke characters named "Jergon McDerp" and "boringspaceshipgame." Even during a climactic battle, *EVE's* bizarre brand of humanity always shines through if you know where to look.

The Northern Coalition fleet began disintegrating immediately, both from the guns of Hellcats, and the flight of pilots who could plainly see that the battle was already lost. The server creaked under the strain of these thousands of ships, struggling to calculate the damage and movements of 4000 pilots relative to each other, while rendering their lasers, ship hulls, explosions, and particle effects.

Many of the Northern Coalition had already begun attempting to log out, and soon the rest were ordered to do so by Imperian himself, hoping to salvage as much from this embarrassing debacle as possible. Some were destroyed when they tried to sneak away in the aftermath of the battle. Over the next two days the true impact of the disastrous situation revealed itself, and as the final straggler ships were caught and shot down the final cost of the slaughter was tabulated.

For the Northern Coalition, this was a mistake that cost a record 10 Titans and a staggering 1.2 trillion ISK, more than double the devastating toll it had inflicted on Pandemic Legion earlier in the war. For the Drone Region Federation, it cost just one-fifth as much.

O2O-2X

The disaster at Uemon left the Northern Coalition skittish and unconfident. If Imperian—the coalition's hero fleet commander who was to deliver them from the Russian threat—could be humbled on that kind of a scale, many wondered what hope the coalition could possibly have. Vuk Lau was having increasing trouble convincing Titan and supercarrier pilots to log in for battles because confidence had been so bruised by the previous losses. The Northern Coalition was said to be full of relatively green industrialists who had managed to earn enough to build or buy themselves a Titan, but really weren't interested in losing their pride and joy in a battle they might not even win.

This was compounded when reports came back from scouts the next day after Uemon reporting that the exact same Russian players whose Titans had been destroyed in Uemon were spotted again at a skirmish in fresh Titan hulls. The belief was that the extremely wealthy Russians had so much liquid ISK at this point that they were able to simply purchase new Titans off the open market. Some of their Titans, it was rumored, were actually purchased from Northern Coalition industrialists who weren't required to do background checks on their clients. It's very difficult to prove whether those transactions took place, but one of the most enduring characterizations of this conflict was the irony of the Northern Coalition selling Titans for profit to a hostile power bloc. In effect, the chaotic Northern Coalition was profiteering off a war predicated on its own destruction.

You could scarcely imagine a more ironic fate, and perhaps that's also why we should distrust that temptingly pure schadenfreude. True stories are not usually so easy to distill into delicious irony.

The replenishment of the Russian supercapital fleet was a burdensome reality, however. As the Northern Coalition's fleet continuously diminished, the Russian fleet seemed miraculously unscathed, growing at a faster rate than the Northern Coalition could hack it down.

After the gaffe at Uemon, the Russian invasion rolled forth into the Northern Coalition region of Geminate as the Russians began to make real progress for the first time.

Imperian tried once again to stymie their progress, and drew a line in the sand at O2O-2X. He had disappeared for two weeks after Uemon, and the Northern Coalition breathed a sigh of relief that its legendary commander was still in the fight. The Northern Coalition had three primary Fleet Commanders: Imperian, Yaay, and Sala Cameron. The loss of any one of those three would have been a wound.

For two weeks, the two coalitions brought their full weight to bear against O2O-2X in three major rounds of battle. In Round 1, on February 26, Imperian led 500 pilots organized according to a new fleet strategy that they hoped would turn things around. They were flying a unified fleet in the "Alpha Maelstrom" style. Essentially, the goal with this fleet was to use well-coordinated long-range battleships to focus-fire on an enemy ship and destroy it in a single volley. Imperian's adoption of this fleet style was partially based on his experience in Uemon, as he attempted to adapt to the fact that "soul-crushing lag" was going to be a reality in this war. The Alpha Maelstrom doctrine helped mitigate the problem of lag because a ship could usually be destroyed in one shot without much room for latency to disrupt the plan.

The 500-ship fleet met a DRF fleet of similar size and soon became mired in stalemate. Eventually they were rescued by three supercarriers from a group of Rebellion Alliance holdouts calling themselves "Gypsy Band," (the word "gypsy" is an exonym referring to the Romani people of Europe and is usually considered a slur) who had stayed together to defend the Northern Coalition after Daroh allowed the DRF across the border. With its help, Imperian cracked the DRF defense, and propped up the Northern Coalition's confidence.

In Round 2, the DRF countered Imperian's tight Alpha Maelstrom formations with masterful bombing runs by its legendary Bomber Fleet Commander "Old Hroft," who was known for his skillful use of area-of-effect ships that could take out dozens of enemy ships at once if properly positioned. Northern Coalition forces were driven back, and the two sides prepared for the next in a series of attacks and counter attacks that might stretch on indefinitely. As so often happens in *EVE*, however, the conflict would not be decided solely by fleet tactics.

YAAY'S DECEPTION

Round 3 of the fighting in O2O-2X came on March 17, 2011, and the Northern Coalition envisioned this as a fight to end the stalemate and force the Russians back into the Drone Regions. The Northern Coalition coordinated its supercapital pilots for a massive showing, and on the day of battle, Yaay jumped into O2O-2X at the head of 20 Titans and 92 motherships, one of the greatest fleets ever assembled in its day. The showing was so large, in fact, that neither Pandemic Legion nor the DRF would dare to engage. However, Yaay didn't want to waste a battlefield advantage by letting the enemy disengage while he had all these ships ready at his command. With no way to force Pandemic Legion and the Russians to engage, Yaay set to work on a plan to trick them into confidence.

 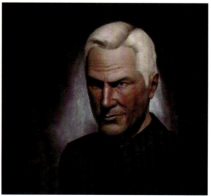 

Above: The wrecks of half a dozen dreadnoughts litter the area as two more are targeted down.
Left Bottom: Pandemic Legion lead fleet command team Elise Randolph, Shadoo, and Shamis Orzoz.

The Northern Coalition knew very well that its alliance was infiltrated by Pandemic Legion spies. They simply weren't well-coordinated enough to be able to root them out and expel them. Who has time for a mole hunt in the middle of a war? These things were just a fact of life that had to be endured. Pandemic Legion was going to find out the Northern Coalition's battle plan, and there was little that could be done to stop it.

Instead, Yaay decided to use his enemies' spies to his advantage. He declared openly on fleet comms that the enemy had retreated and the need for supercapitals was over for the day. He said those ships should leave the system while the subcapitals should stay put and guard the retreat. Pandemic Legion's spies' ears perked up at that order. Yaay had just given a command that would leave his entire subcapital fleet undefended and outnumbered. The spies relayed the orders to the Pandemic Legion fleet commander, who began planning to jump in and destroy the subcapital fleet once the Titans and supercarriers had jumped out of the system.

But secretly, Yaay sent personal communications to his Titan and supercarrier pilots to stay put and ignore the public order.

"On comms Yaay told all supercapital ships and regular capitals to jump out, except several carriers that had cyno-sural field modules fitted. Apparently at the same time on a different channel Yaay told the supercapitals to stay put.

After jumping I loaded grid and heard [the fleet commanders] starting to shout on comms for bubbles, and for all the carriers to light their cynos. About 12 cynos went up and our whole capital fleet jumped back in. Lag wasn't horrible to start, but got bad shortly as their caps and supercaps landed on us.

They apparently thought they were warping on top of a bunch of stragglers, hoping to grab a few easy kills before we were gone. Our hictors and dictors (Heavy Interdictors and Interdictors, anti-warp ships) went to work and we started slowly taking out their Titans."

— Anonymous Northern Coalition (Mostly Harmless) pilot, quoted by EVENews24.com

March 18, 2011

The DRF fleet warped into O2O-2X expecting to meet a few hundred subcapital battleships but instead met the full brunt of the Northern Coalition's power, unmoved and waiting for their enemy. More than 100 supercapitals, one of the mightiest fleets ever assembled in *EVE*, poured firepower into the DRF's most valuable hulls. What followed was one of the most one-sided massacres of the war. Before they managed to escape, the combined DRF/Pandemic Legion fleet lost 18 supercapitals, amounting to 1 trillion ISK, nearly cancelling

Siezing the Council

On March 26, 2011 the results of the sixth Council of Stellar Management were announced to the community. The council had grown in reputation over the past few years as players campaigned in-game and on the forums for the opportunity to represent the vox populi in discussions with CCP.

This time the election saw a record number of votes, more than 10,000 more than the year prior. When the ballots were counted The Mittani emerged as the clear winner with 5,365 votes. The Chairmanship of the Council was handed off from a low-sec pirate warlord named Mynxee (the leader of an all-women pirate gang called Hellcats.) The Mittani was open about what it took to win this victory. Just days before the election he wrote:

"The contest for CSM6 is turning into the biggest all-out electoral slugfest yet. In New Eden, this shining example of 'representative democracy' is prosecuted through an election where lying, cheating, scamming, vote-buying and disproportionate representation of the wealthy is explicitly condoned; in internet spaceship elections, anything goes, as long as it doesn't violate the EULA. And since this is *EVE*, that means that the contest for CSM6 isn't an election—it's a war."

out the Northern Coalition's historic losses in Uemon the month earlier. Such comparisons are never perfect, but the grey market value of that loss could've been as much as €60,000.

"The Solar Legion of Red Noise, [DRF] was expecting the Northern Coalition to use their usual tactics and blob the O2O-2X system to reinforce the station and stay there. [...]

The main mistake was made by a scout, which informed the FCs that NC supercaps left the field. After, the DRF fleet jumped into the system. First DRF Ragnarok was down in seconds (took eight Doomsday hits) by NC Titans before he even managed to enable [shield] hardeners. The usual lag began and half of DRF fleet didn't manage to load in to the system. Some of the supercap pilots lost their connections and didn't manage to get back in-game during the battle. After the situation was clarified, DRF FCs gave the command to supercaps to leave the battlefield. Some of them managed to escape.

Overall lag situation affected both fleets, however, NC managed somehow to bring more people into the system and because they were in the system first, looks like they had little bit less problems with lag."
— Anonymous DRF Pilot
Mar 18, 2011

Though it may seem that server randomness plays an overly heavy hand in this story, it's important to note that the players all knew this very well, and the fleet commanders literally studied the server architecture to learn how it would react to certain pressures.

"The capital ship fleet commander (Yaay) pulled a rabbit out of a hat and showed the Russians that having spies doesn't always pay off," said an anonymous Northern Coalition member quoted by EVENews24.com.

Victories like this helped prop up the Northern Coalition's morale at a precarious moment. Their hero fleet commanders were performing as well as anybody could possibly hope, given that wars on this scale were nothing anybody could prepare for. Nobody actually had experience with campaigns on this scale, the players at the top level of *EVE* have always been exploring the game's rules day-by-day.

The Russians, miraculously, were undaunted. In battle after battle they had lost Titan after Titan, and yet the pressure from DRF supercapitals wouldn't let up. The Russians regrouped, and forced Round 4 of the ongoing, three-week-long battle for O2O-2X. Both sides recognized this was certain to be the defining bout of the fight, and it was all-hands-on-deck for the final battle for the Geminate region.

A familiar sight played out as four Northern Coalition fleets (900 pilots) jumped into the system near a DRF force (1000 pilots) positioned close to one of the DRF's defensive starbases. Both sides waited to find out whose fleet would be more adversely affected by server latency, neither wanting to recommit old mistakes by jumping into a lag-induced disaster.

The server was performing as well as either side could have hoped at this stage, and the two fleets engaged in one of the bloodiest contests of the war, this time trading blows with fully operational Titan Doomsday weapons. The battle remained close throughout, but the scales began to tip when the DRF ships began turning around and "slowboating" (burning sub-light engines) their way inside the shield of a nearby starbase they had erected here earlier. This maneuver allowed their shields to absorb a great deal of the NC fleet's damage output while the ships escaped ultimate destruction, an impressive feat of fleet micromanagement.

In the ever-important Titan count, the Northern Coalition lost six more Titans to the DRF's three. The Northern Coalition was left to absorb yet another near-trillion ISK loss as O2O-2X was finally captured by the Russians.

While everyone on both sides vocally agreed that this slugfest was extremely fun, within the Northern Coalition a certain sense of doom began to propagate.

NCDOT

Complicating matters, two new enemies declared war on the Northern Coalition, and they require a short bit of explanation. When IT Alliance splintered months earlier, its fragments allied with different established groups and formed new alliances. The first group consisted of some ex-IT/BoB corporations working with an ancient enemy of the Northern Coalition. This new alliance had one thing above all else in common: they hated the Northern Coalition.

So deep was the enmity within this group of pilots for the Northern Coalition that they chose a new name for themselves: "Northern Coalition." That's not a typo. There's a little tiny extra period at the end. (Pron: "Northern Coalition Dot".) For our purposes we'll refer to the group as "NCdot."

Two weeks after the showdown in O2O-2X, EVEOnline.com published an article about the mess the Northern Coalition now found itself in titled "NORTHERN COALITION ATTACKED ON ALL FRONTS."

"The Northern Coalition is currently being pressed hard by fleets from the Drone Region Federation, Pandemic Legion, and the [NCdot] Alliance," reads the article by author Svarthol. "Today, a fleet from the [NCdot] Alliance [along with many allies] reinforced five out of eleven Capital Ship Assembly Array towers in the 6OYQ-Z system in [the Tenal region] during a three-hour joint operation. They have also been attacking in Pure Blind and Fade regions. Simultaneously, Raiden (a different group of ex-IT/BoB corporations) and the Drone Region Federation have been attacking towers in Vale of the Silent, Venal, and Geminate. The plan seems to be to collectively deprive the NC of a part of their capital ship production facilities as well as the extremely valuable technetium deposits in the Venal region."

The article quotes a spokesperson for the Drone Region Federation who remarked, "Everything they do is to split Northern Coalition forces. Just like in any war, you need someone making the front line as big as you can. This tactic has always worked against those who can only fight when they outnumber, but if they have split interests and split areas, they will either have to choose or break down."

This bit of strategy proved remarkably prescient. The alliance of forces united behind Pandemic Legion and the DRF was tearing the Northern Coalition apart by the seams. By putting pressure on a large swath of the Northern Coalition's territory, it ensured that dozens of different groups within the coalition had different priorities. One corporation might want to protect their in-build Titan under attack by Pandemic Legion in a vulnerable Capital Ship Assembly Array. Another corporation could be pinched because NCdot had shut down its technetium moon and it needed it operating normally to pay its debts. Meanwhile, more constituents were calling for help on the front line near the border every day.

All the while, Death fomented dissent within the Northern Coalition by lobbying Goonswarm Federation's Coalition of Deklein to stay in the West and largely out of the fight. Death had been a friend of Goons for years—since before the beginning of the Great War in 2007—and that meant he could go behind Vuk Lau's back and speak with The Mittani and Vile Rat, director of the Corp Diplomatique, directly whenever he wanted.

"We were real life friends with Death," said Goonswarm's DaBigRedBoat. "[Ex-Goonswarm leader] Darius [JOHNSON] and Death lived within an hour of each other and were real life best friends."

I told DaBigRedBoat about a rumor I had heard that Death would wine-and-dine potential partners at a local New York City bathhouse, mostly expecting him to laugh it off or demur. To my surprise he instantly exclaimed:

"I've been to that bathhouse [with Death]! It's a Russian bathhouse and you feel like you're in the mafia when you walk in. Like, I thought Death was mafia because he would snap his fingers and servants came out it was hilarious."

Boat added that Death was somebody Goonswarm wanted to stay on the good side of because he made for an unusual adversary.

"Death did something that not a lot of people did," said DaBigRedBoat. "Today it's commonplace for people to multibox and run multiple accounts. Death ran 89. He kept all of it on a spreadsheet. That's how he kept track of it. He would run four Titans by himself, and he would defend his own sovereignty by himself."

THUNDERCATS

Throughout April 2011, the war continued to go badly for the Northern Coalition. The battle reports from this time are full of drastic Northern Coalition losses at every level, from battlecruisers up to Titans. The situation worsened over the course of the month, reaching its zenith on April 27 as hundreds of Northern Coalition pilots fell to Pandemic Legion's brand new fleet concept: Thundercats. Another evolution of its original concept, Thundercats used the Tengu hull with afterburner boosters allowing

Opposite: The Tengu-class strategic cruiser became one of the most popular ships of this era thanks largely to Pandemic Legion's innovations and its overall simplicity. Below: An entire fleet of Tengus orbit its fleet commander in unison.

them to launch barrages of missiles while maintaining a speed so high that enemies couldn't target them.

"I am completely in awe of [this fleet]," wrote the player Ripard Teg, a well-known *EVE Online* blogger, about the Thundercat fleet style. "They have no holes, no exploitable weaknesses... nothing. A fleet of children... a fleet of complete noobs... could get into this ship and win battles, as long as those child noobs had [a lot of skill points] each and could follow a fleet commander's orders."

The combined DRF/NCDOT force was consistently dominating the battlefield, and it was getting worse all the time. To deal with the mass numbers of Northern Coalition pilots, Pandemic Legion had invented a new type of Titan weapon configuration. Previous Titans were fitted with components that upgraded their Doomsday weapons and armor plating to help them survive opposing Doomsdays. Titans had many other weapon capabilities, but these were considered to be a secondary concern to the main Doomsday weapon. But Pandemic Legion's new "Tracking Titans" used rows of laser batteries as well. Ordinarily these huge guns would move too slowly to be able to lock and shoot down a battleship, which is roughly 1/1000th the size of the Titan. But Pandemic Legion's fleet wizards managed to figure out how to compensate for that problem by fitting just the right components with just the right fleet buffs. This allowed the Tracking Titans to one-shot almost anything. Larger ships would fall to its Doomsday while smaller more nimble ships would be mowed down by rows of laser cannons.

Against both the Thundercat fleets and these new Tracking Titans, there was no hope of victory in everyday engagements. The best strategy the Northern Coalition ever figured out for fighting back was its Alpha Maelstrom fleets, usually staffed by Goons and TEST. But these also had a nasty habit of being blown up by the bombers of Old Hroft.

Losses like this were bad for coalition morale. In particular, the western alliances of the Northern Coalition

led by Goonswarm Federation and TEST began to feel that they were sacrificing more in this war than the eastern alliances who started all of this mess. Though The Mittani had always doubted the wisdom in provoking the Russian bear, it had happened anyway. He now had to seriously face the prospect of a future in which half of his coalition was conquered, and the Deklein Coalition would be alone against this massive and growing coalition of enemies. Worse, he saw in the Northern Coalition precisely what he had seen in dozens of other alliances he had spied on when they were in their death throes.

There may be no other person in *EVE* who has such intimate knowledge of what organizations are truly like when they're falling apart. After all, The Mittani had personally engineered several such collapses, and been an avid reader of his enemies' forums during the process. Nobody was more familiar with the predictable pattern of cracks that form within a group under stress. The Mittani seems to have begun to recognize signs of imminent destruction and collapse within the Northern Coalition. It's around this time that he made a series of public statements disparaging democratic institutions like the Northern Coalition in favor of "strong" autocratic rule-by-despot.

In a prepared State of the Goonion address, he referred to the Northern Coalition as "a sort of a loose council without any clear lines of authority" adding that Vuk Lau was in charge "in theory."

In the middle of the DRF invasion, The Mittani gave an interview to PC gaming website RockPaperShotgun.com and was quoted saying:

"Autocracy is the most effective form of government in nullsec. Council systems don't work very well. Goonswarm is very lucky in that we have one large corporation, GoonWaffe, which used to be GoonFleet, which is mostly Something Awful members and has over 2,000 people. Since I'm the CEO of that corporation all the other ancillary corporations in the alliance are relatively powerless, and that works towards an autocracy. [...] Democracy is death. In a situation where

STRATEGIC CRUISER // CALDARI STATE
MODULAR DESIGN
TENGU

WARP BUBBLE IMMUNITY

MODULAR HULL DESIGN

you need to be able to respond quickly and with force to strategic problems, invasions or what have you, you can't wait for a vote."

But history also shows that autocracy in *EVE* is fraught with risk. Heavy lies the crown, and the responsibility necessary to run a large-scale *EVE* organization is equal to or greater than a full-time job. And what happens when your leader gets burnt out and the alliance isn't accustomed to functioning without direct leadership? What's more, I am certain The Mittani wasn't about to accept anyone else as that envisioned autocrat. It's probably fair to say that what The Mittani really meant was that *his* autocracy was the best form of government in *EVE*.

All the while, the DRF machine churned forward, deeper into Northern Coalition space and toward Vuk Lau's Fortress Tribute. The majority of *EVE* still believed this space to be impregnable, and the DRF Legion coalition wasn't taking any chances. As the invasion crawled toward Tribute, the Russian commitment to the invasion grew. In April, a guest alliance of the Northern Coalition, "Majesta Empire," saw its capital sacked. The DRF had overwhelmed the system with 34 Titans flanked by a massive wing of 121 supercarriers, likely the largest supercapital force ever assembled in this era.

As the DRF attacks reached Tribute core, Death was still privately lobbying Goonswarm to stay out of the defense of Tribute. He said that the invasion was effectively over, and the "attacks" on Tribute were just a "victory parade" to blow off steam and celebrate. Nonetheless, it would be best if the Goons weren't there for the defense. The Russians posed no threat to the Goons, Death would say perhaps while nude and boiling in a Russian bathhouse, but it would be safer if the rest of the DRF didn't see the Goons defending Tribute. The sight of the collective defense, he said, might goad the DRF into further attacks. Death told him to stay out of it, and he'd convince the other Russians to go home once Tribute fell and the old structure of the north was dismantled.

It's not clear whether The Mittani was taking him seriously, but their diplomatic talks were disrupted in early May when the Goon forums—their hallowed cultural homeland—came under a days-long, sustained Dedicated Denial of Service Attack that sent them into disarray on the eve of the DRF "Victory Parade." As the DDOS attack was still underway, The Mittani gathered his alliance for a speech to let everyone know what was going on and that

the suspect was PsixoZZ Kahi's DRF alliance White Noise, a recording of which survives on Soundcloud.

"This is going to be pretty informal. I just want to talk to people about what the fuck is going on. [...]

We are investigating the people who are responsible for DDOSing us. We're suffering under a botnet of approximately 25,000 IPs. [...] The initial timing of this attack is very suspicious. It occurred only a couple of days before this 'victory parade' in Tribute. Then Razor's Jabber has also come under attack. The timing of things makes me think the parties responsible are in White Noise, but that might not actually be true. [...]

So, who is White Noise? White Noise appeared out of nowhere with a shitload of supercaps a while back. Their leader is the same now as it was then, a guy named PsixoZZ. At some point they ended up with a shit load of outside capital.

It appears—we're investigating this, we're not entirely sure yet—but it appears that one of the major RMT [Real Money Trading] shops that used to be associated with Against ALL Authorities, essentially moved into backing White Noise. They were able to rapidly purchase and acquire a massive supercapital fleet which we suspect is primarily bought. They buy Titans wholesale off of sale orders. They basically started throwing a shit load of isk around that came from nowhere.

Essentially White Noise appears to be a business. Their FCs and their leadership are on payroll. Nync is one of their FCs and he's made no secret about the fact he's on payroll. The reason why the White Noise situation is special is because [Against ALL Authorities] has a grudge and is burning their southern territory in Detorid and Tenerifis. We suspect that White Noise wants to move into the north for real not to just attack the north and then fuck off as in the MAX campaigns. The MAX campaigns were bullshit. [...] White Noise's territory is being burnt behind them, and they may very well *need* to live in Tribute.

What we're fighting against here is a business model. There was the initial investment in buying the supercapital fleet, requiring the ISK to buy that and put their FCs on payrolls. What we suspect—though this is unconfirmed and we're investigating it—is that they hope to take the North, live in the north, and seed the north with renters such as Raiden in Vale of the Silent.

From Tribute they can control and project power throughout the technetium holding areas of the north. And it's just a tremendous amount of real world money. It's a massive amount of profit that is potentially to be had if they're able to do that. They can recoup their investment very easily.

This is why Deklein [Goonswarm's home region] is under threat if Tribute falls. Because Tribute is centrally located, it is essentially the key to the North. If Tribute and Morsus Mihi and thus the North falls, we are going to be facing a rapacious business that isn't going to be content with just sitting there and saying 'we have enough.' Goonswarm's expansion eventually stopped because we had enough to feed and cloth our people in Maelstroms and Titans etc. Businesses have no such level of contentment.

We're trying to source which RMT storefront is associated with White Noise. We don't police RMT, I'm not CCP's cop even though I am the chairman of the CSM. In this case, however, it appears that some RMTers are coming to threaten us personally. Which means that we have to start caring."

Below: A small fleet of battleships in the process of warping through a stargate.

TIME DILATION

In April 2011, CCP Games introduced a new feature to *EVE Online* with the goal of helping facilitate the huge player battles which were happening with increasing frequency and growing in size.

Thousands of players could now be routinely seen in battles around the star cluster. The solution CCP came up with was the Time Dilation feature which literally slowed down time itself whenever enough pilots were gathered in one place.

The more pilots gathered in one place, the more time slowed down until it reached its maximum at 10% of real time. This meant that the real world and the rest of New Eden cruised by at ten times the speed things happened in major battles or player events.

The Mittani's address then struck a remarkably different tone as he attempted to navigate the complex political situation set out before him and addressed one final topic to try to keep this from spiraling into a war against the whole Russian community.

There is one thing I want to talk about though as we get into this which is uh, we are going to be fighting Russians. Mainly we're going to be fighting White Noise which is the driving impetus behind the Drone Russian Federation. [...] This is something that I really want to hammer home because it's something that has fucked up a lot of other alliances over the course of *EVE:*

We have a lot of Russians in Goonswarm. We've been friends with Russians before we were friends with anybody else. We were friends with Russians when they were united under one banner: Red Alliance. We are going to war essentially against a business, an RMT shop in the form of White Noise.

Do not make this fight into a fight against "Russians." Exciting Russian nationalism, uniting them under one banner, ethnic slurs. [That's] one of the stupidest things you can do in *EVE Online*. I watched Lotka Volterra do this. They'd make jokes about Russians, they'd shit on Russians in local [chat]. A lot of the people in the Eastern Northern Coalition, the people in Vale of the Silent from Eastern Europe, Slovakia, Slovenia, whatever, have issues with Russians and they would shit on Russians in local. This is part of why the DRF was so eager to burn them to the ground.

We are at war to defend Tribute, and to fight against an RMT shop. We're not doing this because we like to fight Russians. So make sure you guys figure your shit out."

— **The Mittani**, GoonSwarm CEO in a 'State of the Goonion" address
April 11, 2011.

The speech closed with an obligatory thank you shout out to the myriad allies who were coming along to support Goonswarm, and an encouragement that pilots should be polite and thankful to those pilots who gave up their free time to help.

"The next three days we have [starbases that will be vulnerable,]" The Mittani concluded. "We're going to see whether Death is telling the truth that this is a 'victory parade' or whatever."

The very next day it became clear this was anything but a parade, as serious DRF attacks struck all of those starbases. The "Horn of Goondor" was sounded, and the Deklein Coalition rallied to Tribute to try to hold back the invasion.

The last refuge for the eastern half of the Northern Coalition was Vuk Lau and Morsus Mihi's home—H-W9TY, the Jewel of Nullsec. ●

THE CLUSTERFUCK

"It is purely upon us to choose the fate of ourselves."
— Vuk Lau, Morsus Mihi, Northern Coalition

The Russian invasion rolled on, and throughout mid-May 2011 there was rarely a ray of hope for the Northern Coalition. Despite Goonswarm pledging its assistance, the situation was so bad that the Northern Coalition was now having trouble finding talented fleet commanders willing to sully their reputations with what were sure to be catastrophic losses.

In the midst of this, Pandemic Legion's Shadoo opened a private chat line to Vuk Lau late one night for a talk. After telling Vuk Lau that his coalition's current situation was a result of the Northern Coalition growing "stale sitting on the same shit for fucking 7 years" Vuk Lau took the comment back to his coalition for a rallying cry.

"Now comes the important part," he wrote on the forums. "Starting from Sunday 22nd of May we will stop sucking or at least die trying. I will do the only thing I am good at—delegate stuff and make your pewpew experience enjoyable."

Vuk Lau announced that he had uninstalled his other favorite video game "World of Tanks," giving his account to two teenagers to grind items for him, and that he would be fully focused on *EVE* going forward.

But in the other half of the Northern Coalition—The Deklein Coalition—it was obvious that The Mittani was already planning a new future without the Northern Coalition. On May 26, he published an installment of his popular "Sins of a Solar Spymaster" blog series which detailed his exploits as a shadowy back-channel operative. In a post titled "The Crisis of the Northern Coalition" he wrote at length about the galling flaws in the organization, and the mistakes it had made along the way. He criticized the coalition's diffuse leadership structure, unfocused fleet strategy, and, of course, the way in which they'd managed to piss off the Russians. The column ended on a starkly direct note:

"Speaking at a personal level," The Mittani wrote, "in years past the NC was invaluable in aiding Goonswarm in our war against Band of Brothers, and we owe them a great debt; because of this, we have been committing many fleets to keep Tribute secure. Yet our efforts will be for naught if these issues are not addressed; already there are indications that the Northern Coalition bloc is entering the early stages of a failure cascade. Stern, engaged autocratic leadership is needed to save the day."

Whatever else one might say about The Mittani as an apiring despot, he was certainly clear about his intentions.

On May 30, Vuk reached his breaking point. With Pandemic Legion gank squads swarming Tribute and constant assaults hitting Northern Coalition capital shipyards, Vuk Lau issued a missive to the Northern Coalition. He announced that an all-or-nothing defense must be made in H-W9TY, or else the Northern Coalition must admit that it was a broken organization unable to withstand this enemy. He was fed up with his allies refusing to commit their supercapital ships to the defense.

(Note that English is a fluent but secondary language for Vuk Lau, and when he says "friends-with-benefits" he likely misspoke and meant something akin to "allies of convenience.")

Opposite: A creative screenshot by *EVE* photographer Razorien viewed from the inside of a fleet mashing together.

"As you witnessed, more than a dozen Tribute [Capital Shipyard Assembly Arrays] along with several important towers were reinforced in Tribute core. I doubt we can do shit today (Monday) but if we don't field EVERYTHING we have including ALL SUPERCAPS on Tuesday evening and defend everything that is reinforced, we can be realistic to say that we lost this war. We need to admit to ourselves that the Northern Coalition is broken in several areas, and it's sad that we had to face reality in the worst possible way, but that's life.

What we have—and none can take that from us—are our communities and friendship built between us over the years. At least, after this we know who are our real friends and who are friends-with-benefits.

Anyway at this moment war is not lost YET. As I said Tuesday is "to be or not be" for the Northern Coalition in its current structure, in the true meaning of that expression. It is purely upon us to choose the fate of ourselves."

— Vuk Lau, Leader of Morsus Mihi, Northern Coalition
May 30, 2011

The Northern Coalition leadership posted a well-known video clip of Jean-Luc Picard from Star Trek: First Contact, in which Captain Picard—pushed to his breaking point by the merciless invasion of the Borg—smashes a glass case and gives an impassioned rallying speech.

"We've made too many compromises already, too many retreats," says Picard. "They invade our space, and we fall back. They assimilate entire worlds, and we fall back. Not again! The line must be drawn *here*! This far, no farther! And *I* will make them pay for what they've done!"

Pandemic Legion fleet commander and frequent scribe of battles and conflicts, Elise Randolph, wrote at the time that at this moment, "the gauntlet was thrown."

Below: Since *EVE* takes place in a virtual realm the normal laws of phsyics often don't apply. Seen here, an entire fleet of Titans mashes together creating a dense ball of supercapital ships. Each ship will emerge from this singularity without so much as a scatch to the paint job.

"PL's own Sky Marshall, Phreeze, accepted the challenge," wrote Elise Randolph. "Together with DRF allies NCdot, Merciless, Raiden, and the relentless warriors of [Intrepid Crossing,] [Phreeze] formulated a bold plan. As the server loaded, the rambunctious PL forces amassed with their trademark fear-inducing Thundercat fleet."

Randolph wrote of how the Northern Coalition successfully assembled its blob fleet once again—the blunt-force club which had got them out of a hundred jams before—an excellent and formidable fleet of 900 pilots.

"Fortune favors the bold," Pandemic Legion Sky-marshal Phreeze is said to have written in fleet chat to the Pandemic Legion pilots. "The road to glory is paved with brave souls, and you have enough spirit to pave that road forty times over. We can stand down now and say we got to the brink of glory and nobody would question our fight. Hey, we just about got there – that's something right? Wrong. Nothing succeeds like success. I say we fight our way into the light, make our dreams become their nightmares."

Elise Randolph's description continues: "The NC gang still had cold feet; mistaking bravery for foolishness they unsuccessfully tried to goad the PL forces into giving up their position. The Northern Coalition were at the precipice of redemption, staring fortune squarely in the eyes, and yet they still were reluctant to take the final plunge. Riddled with self doubt, the NC began to convince themselves that they were going to lose everything; it was clear their hearts were not in the fight. Plagued by their indecisiveness and an unwillingness to fight for something, the NC disbanded all 900 members."

Rather than sacrificing the alliance's supercapital fleet in a vain last stand Vuk Lau made the decision to disembark the fleet and give up hope of saving the Northern Coalition.

The last defense of Tribute failed before it even got started, and like other great last stands throughout *EVE* history, Vuk Lau didn't want to ask his members to risk assets he knew the coalition could no longer replace in a vain attempt to stave off destiny.

Over the next week, the Northern Coalition continued to bleed Titans as members in chaos made mistakes and were caught unaware by Pandemic Legion specialist gank teams. The classic signs of collapse were now deeply evident throughout the Northern Coalition, as fault lines which had been widening for months now suddenly sprang open and a sort of Doomsday evacuation from the north began.

The age of the Northern Coalition had come to its end. Vuk Lau and Morsus Mihi, one of the oldest institutions in the *EVE* community, no longer had any credible claim on power. Morsus Mihi soon split apart and the corporations went their separate ways. Two of the largest including Vuk Lau's corporation joined up with the Deklein Coalition, while another faction fled the north entirely. As they retreated, the DRF/Pandemic Legion forces moved in and took H-W9TY. The Eastern Northern Coalition's pets were fleeing in droves. Meanwhile, The Mittani repeated for any who were interested to listen that the collapse of the North was democracy's fault, and a dictator was needed to stem the tide and keep the North safe from the "rapacious business" which now headed a largely-Russian coalition that controlled half of nullsec.

As the mass of Russian forces consolidated the eastern half of the north it at last approached the border of Deklein Coalition territory. Pandemic Legion—still leading the DRF fleet—argued that they should continue the invasion and use this critical momentum to destroy the Deklein Coalition and cut Goonswarm down before it could fill the power vacuum left behind by Morsus Mihi and Razor. But Pandemic Legion's Elise Randolph was surprised to find the Russians had no appetite left for conquest. Unbeknownst to Elise Randolph and Pandemic Legion, The Mittani made a deal with Death: as long as The Mittani didn't attack its new DRF neighbor White Noise then the invasion would stop.

"They made a deal with the Russians behind closed doors," alleged Elise Randolph and confirmed by Death. "After this unstoppable invasion ripped through the rest of the Northern Coalition, the Russians just stopped on the border of Goon territory and went home."

But Pandemic Legion didn't want to waste a golden opportunity. The Goons and the Deklein Coalition as a whole stood to gain immensely from the collapse of Morsus Mihi. Left unchecked, Pandemic Legion feared the Deklein Coalition would form the core of a new and potentially more dangerous northern bloc.

And so, without the knowledge of its Russian allies, Pandemic Legion began planning an operation to capture and "headshot" the Goon capital, the station "Mittanigrad" in VFK-IV.

Unsupported by its Russian allies and deep behind enemy lines, Pandemic Legion planned a daring operation to shut down the system and keep the invasion rolling. Perhaps it was even hoping that a display of strength might convince the Russians to bandwagon and help destroy the Deklein Coalition.

However, the attempt on Mittanigrad was an abject disaster. The operation got off to a halting start as Pandemic Legion had trouble organizing this particular fleet. Far from headshotting the station, Pandemic Legion never reached the station. Instead, a swarm of angry Goons had been rallied to the station's defense, quickly seeing it (correctly) as a direct threat to the coalition's survival.

Instead of blockading Mittanigrad, the Pandemic Legion fleet was instead completely overwhelmed by a classic Goon swarm, and were forced back into their own headquarters station. This was a particular problem because they were staging out of a conquerable station. If they allowed Goonswarm to blockade them inside they could lose everything.

Trapped inside its own station, Pandemic Legion was forced to think fast to figure a way out of the dire situation. There was no way to fight the vastly superior Goon-led force and no chance of reaching a peace deal after the surprise attack.

However, Elise Randolph remembered that old quirk about starbase shield permissions: if the password is changed, the ships inside will be rocketed into deep space. The starbase shield password was changed and the Pandemic Legion ships were fired like bullets past the Goon blockade and into deep space where they escaped, rendezvoused, and retreated.

To Pandemic Legion it was one of the most spectacular "welps" of all-time, and Goonswarm would never forget—regardless of the attack's failure—that Pandemic Legion's goal was to drive them out of existence.

Pandemic Legion, for its part, was forced to admit its plan to capture Mittanigrad was laughably half-baked, and pulled back to reconsider its future.

DENOUEMENT

Some of the remnants of Morsus Mihi abandoned the north altogether to seek refuge elsewhere. It attempted to evict a smaller alliance in the south named Nulli Secunda, deep in former IT Alliance territory on the opposite side of New Eden. Nulli Secunda proved more capable than Morsus Mihi had anticipated, however, and Morsus Mihi left nullsec shortly afterward. Most of Morsus Mihi's main corporations opted to join Against ALL Authorities to help rebuild that old nullsec stalwart as it recaptured the deep south from White Noise.

With Pandemic Legion now officially repulsed and the Russians heading back to the Drone Regions, The Mittani, Goonswarm, and the Deklein Coalition faced an altogether new dynamic in the north of nullsec. They now found themselves in sole control of their own independent coalition without the oversight of Vuk Lau and the old Northern Coalition council. But their new neighbors were White Noise and Raiden (the latter of which was full of old BoB members,) both of which had deep grudges against the Deklein Coalition.

On the upside, however, a number of splinter groups from the old Northern Coalition were now joining the Deklein Coalition in droves. The Deklein Coalition became a vast mess of various corporations and alliances lacking coordination or coherent ideology: exactly the kind of loose coalition The Mittani had long sought to avoid.

With all of these new recruits flowing in, the Deklein Coalition was appropriately renamed "The ClusterFuck Coalition." ●

Above: Hundreds of players mass in two great balls of ships near a stargate in the far backround in the top left. Below: The engines of a derelict ship scattered into pieces.

WINNING EVE

"There was no animosity. Everyone was just sad. I think
the sadness came from us wanting him to tell us its over.
Please, just tell us its done. Tell us we're finished."
— Sort Dragon, IT Alliance

In a testament to the human-driven nature of *EVE*, a massive change in the *EVE* political state can only be explained with a love story.

In November 2010, the real life player behind SirMolle was busy packing up his real world home and selling off much of his belongings. After eight years sailing the skies of New Eden, he was now planning a flight through the skies of the North Atlantic, to the United States to be with his new family: a diplomat from the Northern Coalition named Slinktress, and her five teenage daughters.

"Basically, I sold everything I owned in Sweden then I took two bags and went over to the US," he said.

"We lost our space to Pandemic Legion, and that's where I met Molle," said Slinktress, an officer in a small corporation allied to the Northern Coalition. "Molle was coming around and picking on them, and I had a proposition to make. I sat on his TeamSpeak for three days waiting to get his attention."

When Slinktress finally got SirMolle's attention she at first wanted to discuss a partnership, but the two found they had that certain spark, and spent hours talking in private chat rooms away from the memes and petty arguments of the common membership.

"Marriage was not anywhere in the imagination, oh my gosh! We had a common enemy, and I guess I was entertaining," she said adding that SirMolle got a whole bunch of grief from IT Alliance members for going off into a private chatroom with "some character named Slinktress."

"We were both leading large groups of people too," she said. "I liked his leadership style, it was very similar

Opposite: The view alongside the hull of a Wyvern-class supercarrier.

to mine: honesty. And all of the things that I hated about BoB [as a member of the Northern Coalition]...if he had known about it there would have been some head-rolling."

Though ostensibly about war negotiations, these secret TeamSpeak rendezvous were about something altogether more human. The warmongering HVAC repairman from Sweden was falling in love with the enterprising American diplomat. SirMolle had conducted campaigns which had reshaped the geography of New Eden. Slinktress was doing her best to protect a small alliance that had been buffeted by those epoch-making events.

It wasn't widely publicized within *EVE*, but those in the know were well aware by now that SirMolle was "busy IRL." He was in the United States now making a new home, dealing with the daily blur of activity that comes with parenting five children, and the worst problem of all was that his new home had crappy internet service. "After February 2011, Molle didn't have decent internet, so he couldn't really play," said Slinktress. "We tried to keep that on the downlow so our enemies wouldn't get too excited."

The attempt to keep it secret wasn't very successful. Their enemies didn't know SirMolle's real life situation, but their spies could sense his absence from his coalition as his corporation leaders began to bicker without him as a figurehead.

HEROES OF OLD

There was a growing rift between two factions within IT that wanted to take the alliance in different directions. There were the old school BoB corporations who wanted things to largely stay the same, and then there were Finfleet and X13, corporations which wanted to pursue a more

roving, mercenary-based life like the up-and-coming Pandemic Legion. Amid all the bickering and egos, Pandemic Legion appeared one day out-of-the-black to assault IT Alliance's home constellation in Delve. Years later, after some chat logs were leaked to the public, the reason for Pandemic Legion's attack was made clear.

"It came out that Finfleet and X13 hired Pandemic Legion to come shoot Reikoku [one of the founding corporations of both BoB and IT] assets in Delve," said former IT fleet commander Sort Dragon. The ideological rift about how the alliance should be run was tipping toward open civil war because SirMolle wasn't around to bridge the divide. Spies relayed intel back to the ClusterFuck Coalition that two factions had split IT Alliance and would soon come to blows.

After the MAX campaigns, the CFC had already started expanding on the old Northern Coalition jump bridge network to more efficiently transport ships to the border in case of an IT attack. Now that network reached nearly to the doorstep of Fountain. When IT attempted to shut down that jump bridge network, the CFC used that attack to justify a full-on invasion.

SirMolle was checking in on the alliance and keeping tabs on the situation, but he wasn't inside *EVE* like he had been before. He talked them through the crisis late at night on TeamSpeak when he had time.

"What you guys will see is a truckload of spin, propaganda," he was recorded saying one night as the TEST invasion moved across the border and spilled through the Fountain stargate network. "You're going to see people posting about 'you lost all your manpower, you have no quality left.'"

When TEST Alliance attacked PNQY-Y, the bridge into the core of Fountain, SirMolle was in America starting his new life. "The first couple of months I didn't have time for anything *EVE*-wise," he said. "Thanks to that, there were too many egos in IT Alliance, and it just crumbled and fell unto itself."

SirMolle had seen all of this before in Band of Brothers. The creeping malaise. The declining fleet numbers. The egos and the infighting. An alliance that is assembled around one person is like a body intimately intertwined with a strong beating heart, and it makes no sense to consider the survival of one without the other. There may be other important organs, but it's uncommon that one can step into the central role of pumping life into the collective.

"IT was full of massive egos," said Sort Dragon. "Myself included. You had the original BoB corps who were the major egos. The actual downfall of IT came down to the fact that the only reason why IT survived was that Molle was always there to counteract the egos. You had technically a ticking time bomb, but nobody expected it to go off because nobody expected Molle to level up and escape *EVE*."

"Molle deserved this, he deserved to have his real life at this point," said Sort Dragon. "He deserved to be happy. Unfortunately, the problem that happens these days with leaders is that when it's time to call it quits and either leave or hand over [leadership,] the leaders are sometimes too arrogant and end up taking the alliance or coalition down with them. Even at the bitter end, the final meeting between all of the CEOs and directors and Molle. There were deals made about the splitting of assets, and Molle still only gave up a small percentage of it. He just left. But he didn't even just leave. He existed without existing. He wasn't dealing with the issues that were coming to pass. So when the mini-civil war happened between [Reikoku] and Finfleet, he could've stepped in at any moment and shut it down but he didn't. And there was a lot of animosity."

When he was able to be around, SirMolle instead focused all his efforts on keeping everyone's nerve steady. He commanded them to batten down the hatches and prepare to hold their headquarters in Fountain, 6VDT-H.

"The frontline is going to be Fountain," he was recorded telling his pilots. "It's not going to be in Delve. It's not going to be Querious. It's going to be Fountain. If someone ran round like a headless chicken and moved their stuff out of 6VDT-H: Get. It. Back."

SirMolle ordered the entire alliance's ships stockpiled in 6VDT-H where the defense would be headquartered. IT Alliance prepared for a long siege, expecting the incoming TEST and Deklein Coalition allies to grind down the vast territory slowly. However, Vile Rat's spies in IT Alliance relayed news about the stockpile, and The Mittani saw an opening to end the war in one stroke.

"Mittens (The Mittani) basically had us where he wanted us because we were too badly in-fighting," said Sort Dragon. "Then Mittens moved in. We realized they had a spy in Finfleet and they had a spy in Reikoku, and they could see the turmoil that was coming. When they realized that X13 and Finfleet were leaning toward not helping in 6VDT-H, he made the call to go for it."

The massive attack on the station eventually evolved into what's commonly known in *EVE* parlance as a "hellcamp," essentially a mass of ships that amounts to an unbreakable blockade. More than 800 CFC pilots formed a blob around the station at 6VDT-H, killing anything that tried to go in or out.

"When the big fight at 6VDT-H happened and we were hellcamped [...] we weren't able to deal with it because Finfleet and X13 pulled their supercapitals out

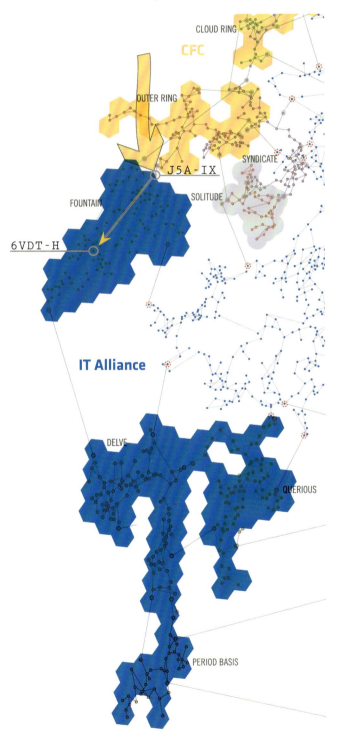

and refused to engage," said Sort Dragon. "I was one of the main FCs that was leading the defense of that. I led most of that Fountain campaign, myself and a guy named Hawk Firebird."

When the final battle for the station came, it was little more than a distraction so that freighters could sneak out as many IT Alliance assets as possible while the TEST/Deklein Coalition fleet was engaged.

"This is/was IT's staging system for their defence of Fountain," said Devilish Ledoux, a Goonswarm member, on the day of the battle, recorded by a forum reporter. "Rather than continuing to degrade their systems over time, we decided to strike at the heart of their strength in the region. It's not a very subtle strategy, but so far, it's been very effective. The fight going on right this moment is their first serious attempt at a defence, [IT Alliance] appear to be much more interested in escape than actual defence at this stage."

"It seems TEST/Goons/[CFC] have had the upper hand all day," said an anonymously-quoted IT Alliance member. "Especially seeing as they steam rolled through a lot of Fountain overnight it seems. [...] I think IT Alliance will lose 6VDT-H and some more."

Entire corporations of players began leaving en masse as they started to sense that IT Alliance wasn't going to be able to survive this latest attack, and the alliance began to swiftly collapse.

"That basically brought down IT because they then decided to go do their own thing and so the rest of us had to escape on our own," said Sort Dragon. Once again, the major corporations of IT Alliance did not die, but were dispersed and scattered to different corners of New Eden.

Within just two weeks of the beginning of the assault, Fountain was taken, and TEST Alliance was installed as the new owners. Fountain would be fertile soil for the up-and-coming Redditors, and 6VDT-H—the site of their victory over IT Alliance—was its symbolic home, a place they would defend to the last.

The fall of IT Alliance was less of a grand battle than a structural collapse. The great, hastily-constructed ship was sinking before the community's eyes. Once again, one of the pillars of the community, the second largest organization of players in the entire game, simply ceased to exist. This time it wasn't betrayal that dispersed the organization, but rather a complete structural dissolution. The invasion began in late January 2010,

Above: Wreckage from destroyed ships dots the background as another explodes under heavy bombardment.
Opposite: The CFC begins its attack on Fountain at the crossroads system of J5A-IX.

and by mid-February IT Alliance's membership had dwindled from 8,000 at its height to now just 2,500.

"Molle was like the dad of BoB and IT, and you can't talk bad about your dad," said Sort Dragon. "Do you want to be the guy who comes to him [as a newlywed] and says, 'why aren't you leading your space empire?!' There was no animosity [toward SirMolle personally.] Everyone was just sad, they weren't mad. It was just an understanding that IT was finished. I think the sadness came from us wanting him to tell us its over. Please, just tell us its done. Tell us we're finished. That was the biggest thing that was needed. Was for Molle to say, 'OK my children it is OK to go.' And that is something that I will take with me for the rest of my life as a leader. Is that when it's time to pull the curtains and it's time to leave do what's right. Close the door. Because your people deserve that kind of honesty."

There are occasions in *EVE* in which players report experiencing a profound awareness that larger events are transpiring. They don't always feel comfortable saying it out loud, because they feel silly being moved by the events on their computer screens. When you've seen these things for yourself, however, you can sense in the way they talk about certain events that there was something truly unique about their experience. People within the game and who are connected to the community can feel things happening even though they're not directly involved. The whisper network of *EVE* is vast and interconnected, and when events like this occur there's a special sense of awe

that comes from knowing that events so much larger than yourself are taking place all around you. In this case the event in question was the rapid evaporation of IT Alliance.

By the time TEST and the Deklein Coalition had taken Fountain, IT Alliance was a functionally broken organization. While the northern half of IT's turf crumbled, the southern end was swiftly invaded as well.

As TEST stepped into its new home both the leadership and the line pilots saw that TEST's time had arrived. Beyond that, a new coalition was organizing in the south to take advantage of the fall of IT. Among them were Pandemic Legion as well as a former IT Alliance renter called Nulli Secunda, "Second to None," led by a player named Gorga and his infamous fleet commander: ProGodLegend.

But SirMolle didn't even mind. In Virginia, SirMolle and Slinktress were building their new life, and SirMolle was preparing for his first alliance barbecue since moving to America. When a player uses the lessons and relationships they build within *EVE* to eventually quit the game and move on to a new phase of their lives, they are often congratulated by the community for "winning *EVE*."

"[SirMolle] claimed in one article that because he got me, he won *EVE*," said Slinktress. "And I felt the same way. My archnemesis was [redacted] and once she found out Molle was marrying me she up and quit the game. I guess we're romantic." ●

THE SUMMER OF RAGE

"This [statue] was once a memorial to the winners of a riddle contest sponsored by late entrepreneur Ruevo Aram. After standing proud for half a decade, it was destroyed in late [2011] by capsuleers who were staging a mass uprising against an intolerable status quo of intergalactic affairs. Today, the ruins of this once-great work of art stand as a testament to the fact that change is the universe's only constant."
— Plaque on the remains of a statue outside Jita 4-4

During the tempest of the DRF/Northern Coalition War, *EVE Online* was reaching the peak of a quiet ascension. The number of players subscribed and involved in the community had been growing steadily since 2003, but that pace had quickened dramatically in 2005 and kept up at a brisk pace of growth through to 2010, when the persistent rise finally began to reach a high plateau. *EVE Online* made global news again as its population grew past that of Iceland, its mother nation.

The main metric the community uses to gauge the health of the game is called "Average Concurrent Users" which is the average of how many accounts are logged-in to *EVE Online* at any given time throughout a month. In January 2011, *EVE* reached its highest ever ACU—roughly 60,000 people online in New Eden at any given time—and the company was in the midst of a celebration of its vision for *EVE.*

The main player hub in *EVE* for that massive concentration of players was the trading hub "Jita" or more specifically, the Caldari Navy Assembly Plant orbiting the fourth moon of Jita's fourth planet.

Jita is the most famous system in *EVE Online*, and has been essentially since the game's founding. Jita is a place unlike any other in *EVE*. It was organically chosen by the players to become the market hub of *EVE Online* due entirely to its natural geography. It's a high-security

Left: A statue of an old man gesturing to the stars that stands in Jita near the most populous player hub station. **Above**: The logo for the much-hyped Incarna expansion patch.

system that is near nullsec and lowsec. It's also home to a space station owned by one of the game's NPC factions, where players can buy and sell goods.

Perhaps appropriately, Jita is also utterly reviled, even by *EVE's* own players. It's the center of New Eden's industry and commerce, and it's usually the most populous star system. Many—if not most—of New Eden's 7500 star systems are often nearly vacant, but Jita's status as the player market hub means it is routinely filled with thousands of players. Precisely because of that, it's also absolutely rife with scammers and bots. All of that traffic, scheming, and harassment makes it into a place that is bustling, thriving, conniving, and singularly despised. In other words, no other place is quite so quintessentially *EVE.* Whether its haters like it or not, under Jita's wrist is the pulse of New Eden.

EVE Online was no longer the plucky Icelandic social experiment from 2003. Now it was one of the biggest products in the video game industry, with more than 500,000 subscribers bringing in millions of dollars per month for a growing online gaming company with global ambitions.

CCP Games' ambition to create a virtual space that survived for decades was starting to seem downright practical, and Jita was its capital. The company even began using the phrase "EVE Forever" as a tagline in advertisements, a nod to the mostly-serious idea that *EVE* could become the first virtual space to achieve actual permanence and never be shut down.

As *EVE Online* grew and grew, the player community was united in celebration with CCP. The relationship between CCP and the community had been badly damaged by the T20 debacle of 2007 in which a developer was found to be cheating. But as time rolled on and the game continued to grow, the two sides remembered that their fates were linked, and the players again began to see success for CCP as success for *EVE*.

However, as we will see in this chapter, business decisions made in Reykjavik, Iceland soon spurred mass protests in the Jita star system within *EVE Online*. It all began with a single phrase that launched a mass community rebellion:

"Greed is good?"

AMBITION

Iceland—the home country of CCP Games—was, in the late 2000s, recovering from literally the worst banking crash in the history of economics. To make matters worse, in early 2010, the tiny island's volcano Eyjafjallajokull (pron: eya-fyatla-yoktl) erupted, shouting 750 tons of magma per second into the sky, blanketing the country and half the continent of Europe in ash, and disrupting economies and air travel systems.

When CCP Games looked inside its servers it saw a virtual world that was somehow more financially stable, less volcanic, and if you could believe it, more populous than Iceland.

Not only was *EVE Online* becoming one of the most enviable products in the gaming industry, with the most unique player experiences; from the perspective of CCP, *EVE* looked like

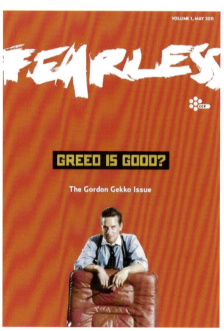

Above: The May 2011 issue of Fearless, an internal CCP Games newsletter which was at the center of community unrest in 2011.

the future of humanity itself. Even as the world literally melted down around them, CCP Games presided over eight years of year-over-year growth for *EVE*.

The core of the company was a group of people whose most outlandish and optimistic idea of the late-90s (*EVE Online*) had not only succeeded but had become a roaring global success worth hundreds of millions of dollars. What do you dream of when you're already living in your dream world? As *EVE*'s player base expanded, so did the developers' vision for the game.

In the late 2000s, CCP dreamed of building not only an entire suite of films and television shows based on the player stories you're reading in this book, but also an entire ecosystem of EVE-based games that linked together into a single virtual multiverse. One day, CCP hoped, players would be able to grab a joystick and fly as a fighter pilot in a first-person starship simulation (*EVE: Valkyrie*) in battles led by fleet commanders playing *EVE Online*, and then dock their ships in their newly conquered space station, meet each other at a bar as their avatars and then streak down from orbit to the surface of a planet to a first-person shooter battle happening on the ground (*DUST 514*) to fight for control of a defense platform which would actually affect another battle somewhere in *EVE*. CCP envisioned a future in which *EVE* organizations were sending strike forces into other video games to better control the ebb and flow of power in *EVE Online*.

It was a daring and expensive vision, and even with *EVE Online* bringing in tens of millions of dollars per year CCP needed to find opportunities to make more money by bringing new services to the players.

The other trend occurring at this time that heavily affects this story is the increasing prevalence of the sale of virtual goods in online games. Prior to ~2010, the vast majority of online video games used one of two business models: 1) the subscription model, which charged users a monthly fee for continued access to the game, and 2) the pay-up-front model which gave users free access to the servers once they'd paid the initial purchase price. *EVE* used both. However, around this time several popular games began to support themselves through the sale of virtual items like avatar clothing and consumable buffs.

Gaming communities—where vicious arguments over game balance are basically constant—worried that greedy corporations wanted to sell gameplay advantages to rich players. In the worst cases, it was obvious that certain companies had used gameplay design itself as a subtle psychological hook to keep players paying up.

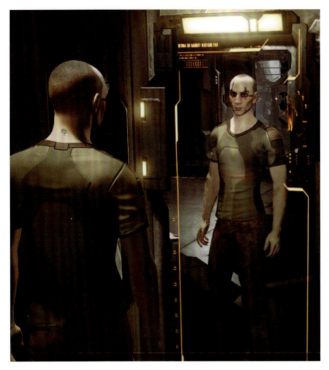

Above: *EVE Online*'s Incarna expansion was an attempt to bridge the gap between players and their ships by embodying each player as an avatar with a full body rather than simply a small square character portrait.

Inside CCP Games, a debate arose about how this new model might interact with *EVE*. Like a lot of large companies, CCP published a company-wide newsletter to help keep the company's employees on the same page. Called "Fearless," this newsletter was a platform for company debates and impassioned op-eds advocating for changes in strategy. In the May 2011 issue, Fearless featured a striking, deep red cover page adorned with nothing but the image of ultra-capitalist villain Gordan Gekko propped up against his signature 80's leather power chair. Above his head read the bolded words "Greed is Good?"

The theme of the issue was the monetization of virtual goods, and the Letter from the Editor kicked off the discussion:

"As *EVE* edges closer to being the grand dame of gaming, turning 8 years old this month, and our other titles continue their prodigious growth our development roadmap is shaping up stronger and better," reads the May 2011 issue of Fearless.

"However," the editorial continues, "as a subscription-based Golden Goose, *EVE* needs to incorporate the virtual goods sales model to allow for further revenue—revenue to fund our other titles, revenue for its developer: you."

What was being proposed was that *EVE Online* should become one of the only games in the world to use all three major online gaming business models at the same time. Not only did *EVE Online* cost money to download, it also cost a $15 monthly fee, and the company now wanted to introduce a new series of cosmetic character items that were gated behind a paywall.

All eyes focused on the impending release of the next *EVE Online* expansion which was the thesis statement for CCP's vision for the future of *EVE*: "Incarna," from the Latin for "flesh."

WORLD OF DARKNESS

The Incarna expansion was to be the fulfillment of a longstanding dream of CCP Games. Because *EVE Online* is famously difficult for new players to grasp, the company had long sought to make *EVE Online* more accessible and understandable to the average person. One of the big problems, they believed, was that average gamers don't want to play as a spaceship. CCP believed they wanted to play as an avatar who pilots a spaceship. It was a subtle semantic difference with enormous design and production implications.

Their solution and vision for the future of *EVE Online* was the "walking in stations" feature. Previously, *EVE* players' avatars were little more than a small picture in the corner of their user interface. With Incarna, CCP spoke of a dream for *EVE* in which players could dock their ships and walk around a personal space called their Captain's Quarters, or even "ambulate" around major stations like Jita 4-4 and encounter other players at shops, bars, and meeting places. In the most far-flung visions, it might even be possible for players and mercenaries to assassinate one another in these public spaces.

"Although *EVE Online* was CCP's flagship product, the company was also in development of an avatar-based vampire MMO known as *World of Darkness*, as well as a ground-based first-person shooter known as *DUST 514*," wrote *EVE* journalist Matterall in a retrospective. "Rather than use an existing game engine, CCP began to develop its own proprietary graphics engine known as Carbon to power these avatar-based games. Carbon would also be the engine used to develop *EVE*'s planned expansion into avatar-based game play."

The development of Incarna takes place parallel to many of the events of this book, and had been unfolding for years already. It was first announced all the way back in 2007—four years prior—and had become just another promised feature on the development backlog.

As the expansion edged closer to release, CCP Games journeyed to the Electronic Entertainment Expo—at the time the biggest event of the year in gaming—to reveal its multi-game vision for *EVE*, focusing on the first public information about its first-person shooter game *DUST 514*.

However, there was a perplexing detail in their presentation that caught the *EVE* community completely off-guard, and it would eventually cascade into a full-on public relations crisis that forced CCP Games to layoff one in five employees:

CCP announced that *DUST 514* would be a console exclusive. Available only on Sony's Playstation 3 platform.

PLAY B3YOND

Many *EVE* players saw this as an open backstab, given that *EVE* had always been a PC-only experience. They saw a nakedly ambitious move by CCP to try to draw millions of "console gamers" into *EVE Online*. Many in the *EVE* community were perplexed and enraged to find that the game intended to expand their community's experience was in fact not playable by a large portion of them who didn't own the console.

Thousands of *EVE* players voiced their discontent on the forums and early social media...all at once. It happened so quickly that at first CCP didn't fully understand what it was dealing with: the opening stages of a full-on community revolt. CCP thought these were run-of-the-mill forum controversies that would run out of gas in a couple of days.

If you look at the Average Concurrent Users for *EVE Online* there is almost always a boost in player activity after a new expansion debuts as players reactivate their accounts to experience all the new stuff with their friends. But after Incarna there was a net *decrease* as disappointment gripped the community instantly.

The expansion which had consumed years of *EVE Online* development time was a massive disappointment. Promised features were missing, and the ones that were included often came with ironic catches. When players docked their ships in a station they would now appear inside the station as an avatar. Except the space you could move around in was a static room of only about 10x15 feet, and was skinned as a rusty Minmatar (one of the game's lore races) station no matter where you were in space. It also lagged badly, caused equipment failures in some players' machines, and inherently removed many *EVE* players' favorite pastime: looking at their ships in their hangar. Now, instead of looking at their beautiful ships they were cursed to the inside of a bland, lonely, rusty metal prison cell populated only by an avatar they had never seen before and thus had no emotional connection to.

The other feature that had launched in the first wave of Incarna was the Noble Exchange or "NeX Store" which was essentially a digital shop for players to augment their new player avatar with vanity cosmetic items like jackets, boots, sunglasses, and a curious little item that ended up making enormous waves: a metal monocle.

The cybernetic monocle could be purchased with a new in-game currency called "Aurum," but when players did the math they realized that the going rate for the eyepiece when translated to USD was an astounding $70. In modern times we know these are textbook conditions for an internet community revolt, but CCP was ahead of its time, and did not yet have the benefit of history. The stage was set for what would become "Monoclegate."

"It didn't take long for people to realise that something was fundamentally wrong with the prices on the Noble Exchange," wrote Brendon Drain for Massively.com, a publication that covers online role-playing games. "At around $40 for a basic shirt, $25 for boots, and $70 or more for the fabled monocle, items in the Noble Exchange were priced higher than their real-life counterparts."

"Given the fact that player avatars could ambulate no farther than the Captain's Quarters, these overpriced items looked like a cash grab by the company," wrote Matterall in a retrospective. "Overall, Incarna was a colossal disappointment after years of hype and mismanaged player expectations. The mood of the playerbase shifted rapidly from frustration to outright disgust."

The following day the controversy within the community reached an even more fevered pitch. On June 22, amid all the community furor, somebody inside CCP Games leaked the May 2011 issue of Fearless on the backchannel forum Kugutsumen.com.

"THE DAY EVE ONLINE DIED"

The hardcore fan community exploded with fear and skepticism about the future of *EVE Online*, with many leaping to the conclusion that this was a death knell for *EVE* as CCP would inevitably slip into the same shady business practices that had been seen in many other contemporary games.

For the first time, the wider community was able to read along with CCP discussions, and gain an understanding of how the company viewed and discussed issues internally.

The players had little context for the purpose of the corporate newsletter. CCPers say that its purpose was mainly to foster discussion within the company about key issues and at times play devil's advocate to explore controversial points of view, the kind of thing Icelanders take pride in. However, the popular perception abroad was that this was a glimpse into CCP's secret psyche, a greed wart that it had hidden from the players for years and allowed to fester.

Developers tried to soothe fears, but ended up only contributing to the utter disaster that was unfolding. Senior Producer "CCP Zulu" wrote a developer blog to address player complaints.

"This week has seen quite a controversy unfold," Zulu wrote. "In almost the same instant as we deployed Incarna—which by the way is one of our more smooth and successful expansions, not to mention absolutely gorgeous—an internal newsletter with rather controversial topics addressed leaked out. To further compound the confusion there was a clear and rather large gap in virtual goods pricing expectation and reality with a large segment of the community."

"While it's perfectly fine to disagree and attack CCP over policies or actions we take," Zulu continued, "we think it's not cool how individuals that work here have been called out and dragged through the mud due to something they wrote in the internal company newsletter. Seriously, these people were doing their jobs and do not deserve the hate and shitstorm being pointed at them."

Unfortunately, CCP Zulu's impassioned plea to spare the average employee of CCP Games fell on deaf ears as some players instead focused on comments he made in the blog which exacerbated the situation. In particular, the community was ruffled because CCP Zulu compared the $70 digital monocle to a pair of $1000 vanity jeans from a Japanese boutique.

The blog closed with a defiant statement that it was likely CCP would later introduce a variety of items for sale that were both more affordable and more expensive than what was currently available.

To make matters much, much worse a defiant all-company email from CCP Games CEO Hilmar Veigar Péturs-son was then leaked online in which he urged employees to ignore the outcry.

```
"sent by hilmar to ccp global list

We live in interesting times; in fact
CCP is the kind of company that if
things get repetitive we instinctively
crank it up a notch. That, we certain-
ly have done this week. First off we
have Incarna, an amazing technologi-
cal and artistic achievement. A vi-
sion from years ago realized to a point
that no one could have [imagined] but
a few months ago. It rolls out without
a hitch, is in some cases faster than
what we had before, this is the pin-
nacle of professional achievement. For
all the noise in the channel we should
all stand proud, years from now this is
what people will remember.

But we have done more, not only have
we redefined the production quality one
can apply to virtual worlds with the
beautiful Incarna but we have also de-
fined what it really means to make vir-
tual reality more meaningful than real
life when it comes to launching our new
virtual goods currency, Aurum.

Naturally, we have caught the at-
tention of the world. Only a few weeks
ago we revealed more information about
DUST 514 and now we have done it again
by committing to our core purpose as a
company by redefining assumptions. After
40 hours we have already sold 52 mono-
cles, generating more revenue than any
of the other items in the store. [...]

Currently we are seeing _very pre-
dictable feedback_ on what we are doing.
Having the perspective of having done
```

this for a decade, I can tell you that
this is one of the moments where we
look at what our players do and less of
what they say. Innovation takes time to
set in and the predictable reaction is
always to resist change. [...]

     All that said, I couldn't be prouder
of what we have accomplished as a company, changing the world is hard and we
are doing it as so many times before!
Stay the course, we have done this many
times before."

     — Hilmar Veigar Pétursson, CEO, CCP
     June 23, 2011

Like the straw that broke the camel's back, the *EVE* community collapsed into what felt like open revolt. Within hours pilots began organizing outside the Caldari Navy Assembly Plant in Jita 4-4. Soon after, they got the idea to orbit the famous Jita Memorial statue nearby to better attract attention.

What followed was perhaps the greatest achievement in the history of the oldest of online gaming traditions: the spontaneous conga line. The protestors' massive conga stretched around the years-old statue like a great wheel. The spokes of that wheel were the lasers and missiles they fired at the statue—erected to honor the winners of a riddle contest years earlier. Though it was an aggressive display, the monument itself wasn't a destructible object so the only effect was a beautiful cacophony of colors and particle effects lighting up the skies of Jita. The ordinarily sturdy Jita supercomputer server hardware lagged under the strain as thousands of players undocked to join the great conga line orbiting the stationary 3D model of a robed old man gesturing to the stars.

But players weren't just protesting with their virtual lasers, they were also using their cash. Thousands of players deactivated their alternate accounts in a show of protest. While ordinarily a new expansion should entice thousands more players, instead onlookers were scared off from trying *EVE* at all as this increasingly looked like a virtual world on the brink of collapse. Yet no one could look away, because once again something was happening in *EVE Online* that nobody had ever seen before. Protests had happened in many online worlds by now, but never on this scale, and never with such splendid screenshots.

"The Council of Stellar Management is being flown to Iceland to discuss the issue in an emergency meeting, and I seriously hope that something good comes out of it," wrote *EVE* reporter Brendan Drain in an article on June 26, 2011. "I don't want to look back on this weekend in years to come and say to people, 'This was the day that *EVE Online* died.'"

STELLAR MANAGEMENT

To attempt to quell the controversy, CCP called a special meeting of the player-elected Council of Stellar Management to hear the voices of the players and begin to understand how things had gone so very wrong.

For two days from June 30-July 1, the CSM was in the CCP Reykjavik offices involved in intense negoti-ations. Previous CSMs served mostly at the pleasure of CCP Games, but this time the council knew it had real leverage to not only enforce changes to the game but also to advance the station of the CSM itself and make it a more indispensable institution. The CSM this year was a Who's Who of nullsec leadership featuring not only The Mittani as its Chairman, but also Elise Randolph of Pandemic Legion, Death, Draco Liasa of RAZOR Alliance, and Goonswarm diplomat Vile Rat. At the end of the all-night session both parties agreed that the negotiations had been intense and fraught yet fruitful.

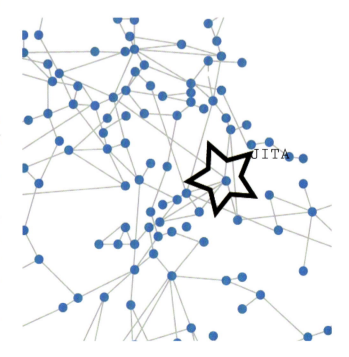

Above: The great conga line surrounds the statue in Jita 4-4. Opposite: The trade hub system of Jita lies in a dense cluster of stargates, and while I doubt this image will do much to help you visualize its location it illustrates why the players would need to choose one system as the understood meeting place.

Even as protesters continued to swarm the monument in Jita 4-4 CCP Zulu appeared in a joint video statement with the Chairman of the CSM (The Mittani) to address the community and hopefully quell the unrest. Taken in context, the resulting video is one of the more fascinating artifacts of *EVE Online*'s history. Two members of the *EVE* community—one who worked his way up through CCP Games, and another who gained power through the virtual environment—sit opposed to one another. Miraculously, the one with power derived entirely from the virtual realm is the one in a commanding position of authority. While The Mittani lounges in his fabric chair in front of the camera—spiky gelled black hair, and a chin capped by a goatee—Zulu sits tensely and leans as far away from The Mittani as possible. Both of them agreed that the negotiations over the past two days had been fraught.

"We were a little mean," The Mittani says at one point in the video. "I'll go ahead and allow that we used strong language. It was actually maybe a little bit awkward when we watched the expressions on the faces of the CCPers as they read our statement."

The statement signed by the Council of Stellar Management reads,

"We believe that the situation that has unfolded in the past week has been a perfect storm of CCP communication failures, poor planning and sheer bad luck. Most of these issues, when dealt with in isolation, were reasonably simple to discuss and resolve, but combined they transformed a series of errors into the most significant crisis the *EVE* community has yet experienced. We hope that this meeting will be the first step in the restoration of trust between CCP and the *EVE* community, and we will keep the community informed as to CCP's efforts in delivering on the commitments they have made to us and to you."

A testy truce was agreed to in Iceland with both parties publicly agreeing that CCP had no plans to abuse pay-to-win business models. However, the meeting and the Jita Riots had exposed a deep vulnerability in the *EVE Online* business model. What happens if the players within the virtual community begin to organize? Is CCP really in charge if the players are willing to cancel their subscriptions en masse? In early online games the slang for "developers" was literally "gods" because they ultimately controlled the off-switch for the game and could change it in any way they saw fit. The Jita Riots had raised a fascinating question: who was in control of the off-switch of *EVE Online*? The early online game devs would often shut down their virtual worlds after time had passed, interest had waned, or the workload became too great. CCP Games no longer had the option of walking away. It had built a company of 600 employees off the success of *EVE*, and now multiple other game production teams were being financed by its success. It's not melodramatic to say that the sudden failure of *EVE* could have had notable consequences for the nation of Iceland. CCP now had investors and employees with children to consider. No person or group of people at CCP was capable of making the choice to shut down the game. The players, however, had shown that they ultimately could, or at least that they could briefly put it into cardiac arrest. Player protests of this sort could affect earnings, and even a hiccup in payments could drastically disrupt operations at CCP.

EVE Online itself had encouraged the players to form these sprawling organizations, and now those very organizations were the grassroots of a movement that had spread across the star cluster and had miraculously proved that the *EVE* community was somehow more in control of *EVE* than CCP Games itself. If the game died, it seemed, it wouldn't be a result of developer intervention. It would be because the players walked away.

"October 5 marked the official end of the Summer of Rage with two devblogs," wrote Matterall—an *EVE* journalist and podcast host of "Talking In Stations"—in a retrospective. "The first was from CCP Zulu who communicated an 'immediate refocusing of all the *EVE* development teams on *EVE*'s core gameplay: spaceships.' *EVE Online*'s Winter 2011 expansion, which would come to be known as Crucible, would include a laundry list of improvements and fixes. The second was from CCP CEO Hilmar Veigar Pétursson who issued his own apology letter admitting, 'I was wrong, and I admit it.' It was a humble end to a period that witnessed a player revolt that cost CCP roughly 8% of its subscriber base. However, it was also the beginning of a new era when CCP would become better communicators, more engaged with the players, and more focused on fixing the issues that had piled up for so long."

CCP Games was left in Iceland with a new reality to contend with. The temperature in the community was finally beginning to normalize and the protests were dispersing. Still, the number of subscriptions to *EVE Online* was *down* 8%. This after hoping that Incarna would usher in a new era of CCP Games in which *EVE Online* would finally become "the grand dame of gaming."

Instead, the community realized its position of power, and CCP Games was forced to face an internal reckoning. With subscriptions declining for the first time in almost ten years and progress slowing on both *DUST 514* and *World of Darkness*, CCP Games announced a 20% staff cut. One in five of CCP Games' 600 employees was to be laid off immediately. It was a traumatic moment for everybody involved in both the community and working at CCP Games as the consequences for the unrest affected developers who had nothing to do with it. Primarily from the *World of Darkness* team in Atlanta, Georgia, but affecting many departments of CCP. Slowly, somberly the game went on.

IN THE RIGHT HANDS

After the Emergency Summit, The Mittani and Vile Rat were back to their indefatigable space villainy, carrying out their latest large-scale galactic exploitation scheme. Their new plan was called the Ice Interdiction, and was a plot to use the mass Goon membership to spread out across all the hundreds of asteroid belts in high security space that produced a valuable material called an Oxygen Isotope, and systematically destroy the players who made their living in *EVE* mining it. The Goon "Dread Pirates" fanned out across space and enforced a ruthless cartel.

"On 9/30/2011 The Mittani crossposted on multiple forums, both public and private, a manifesto addressed to Goons announcing the beginning of the Interdiction," reads the *EVE* wiki about the Ice Interdiction. "Well constructed and wide encompassing, the document covered the reasoning behind targeting Gallente Ice belts, ganking tactics, fittings, reminders of previous commitments in Delve and CONCORD evasion tactics. Summarizing, the stated intent was to bring the global *EVE* economy to its knees by eliminating the playerbase's supply of Oxygen Isotopes, a versatile ice product required to fuel Gallente/ORE jump drive capable ships, player owned stations and a vital ingredient in the Tech 2 production recipe. The desired result of all these efforts was nothing more than engendering ill-will and hatred of Goonswarm by the rest of New Eden's capsuleers."

As 2011 drew to a close, and a new election season loomed, The Mittani mused on the past year.

"Vile Rat and I headed to Iceland in December for Yet Another CSM Summit. The good news is that this one wasn't awful, like the atrocity of the Emergency Summit. The fruits of our efforts in seizing the CSM last March have been made obvious by Crucible; Time Dilation was successfully used on Tranquility just the other day, and I expect it to be turned on permanently soon. That alone will utterly change the nature of warfare in *EVE*, increasing the emphasis on subcapitals and the effectiveness of Logistics ships.

There are perhaps three pubbies left [on the forums] who are still chanting the old 'CSM is Useless' mantra; the rest have actually gotten it through their mong-brains that the CSM matters (in the right hands) and can produce results (in the right hands) - and so have shifted their wailing and ranting to the more pleasing refrain that the CSM has too much power and is a danger to the game, and that (the horror!) goons are on the CSM.

The CSM can wield a frightening level of influence if someone of sound mind and a knack for political manipulation is in charge. [...]

So: we are approaching Election Season once more. Last year we - along with other like-minded members of nullsec - swept the election and helped remake *EVE* into a spaceship game. There will be many more votes this time around, as CSM6 raised the stakes for the entire game. I will once again be running for the Chairmanship, and we will be fully mobilizing to ensure that the voice of ~the people~ (ie, our people, and everyone of like mind to us) is heard. This will be a high-stakes election, not merely because of the power we have created within the CSM, but because after blowing up thousands of miserable Empire barges, I suspect the pubbies don't like us much!"

The Mittani, CEO, Goonswarm Federation
January 14, 2012 ●

DEATH AND THE MASTER

"One alliance leader betrayed the other and got money for
it. This is the exact reason why two bears, living in the
same place, stopped talking to each other. Completely."
— BigMaman, former right hand of Mactep

In the Summer of 2011, the Drone Region Federation was undisputed as the most powerful group in nullsec. It had crushed the old Northern Coalition and forced the ClusterFuck Coalition into a truce. It's alliances controlled more than half the star cluster, but the peace was a testy one held together because everyone was making ISK. Once again, there was no existential threat to Russianness, and it was allowing the community to fray. It still wasn't public knowledge that the Russian founders weren't on speaking terms.

By January 2012, the Drone Region Federation will face a cascading sequence of disasters that tears the coalition apart. The ClusterFuck Coalition will annex its northernmost regions nearly unopposed, a group of enemies the Russians long thought defeated will invade and conquer RA Prime, Red Alliance will collapse, and Mactep will take the Drone Regions from Death by force.

Where to begin?

BROTHERS

Mactep and Death were the most famous Russian players in *EVE* history. The two had stood by each other at the famous Siege of C-J6MT in 2006, which arguably crystallized the entire Russian community and created an origin story for Russiandom in *EVE.*

"We were like brothers," said Death about Mactep. "We were pretty much like brothers."

Across the forums and IRC channels of the early Russian internet, this legend spread far and wide and

Opposite: An NPC drone near a hive in desolate
Drone Region space.

turned Mactep and Death into the Arthur and Lancelot of Russian internet legend. Or perhaps more accurately the legendary bogatyrs (knights) of the steppe: Ilya Muromets and Dobrynya Nikitich. Mactep was the fearsome fleet commander; Death was the cunning politician and industrialist. Legends ran wild on the Russian forums about the improbable victories Mactep achieved on the battlefield, and about his devout best friend Death who ran 89 accounts.

Together, in 2005, they brought the Russian homeland in Insmother back from occupation. Together they destroyed the Coalition of the South, which had been Insmother's occupier. Together they led the Russians of *EVE* into a punitive war against that Coalition's allies and helped destroy the hegemon of early *EVE*: Band of Brothers. Together they swept through the Drone Regions in 2009 and together they recaptured C-J6MT from Bobby Atlas and turned RA Prime into the capital of a dominion that spanned hundreds of star systems. It was a story of friendship that spanned five years, and surely involved some of the most treasured memories of their lives. Space fame and fortune were supposed to be their just reward, and yet just as in real life, fame and fortune can result in an altogether more depressing reality: loneliness and peerlessness.

People close to Mactep say his cosmic notoriety led him on a personal path to ego and authoritarianism as he hoarded power within *EVE* even as he drifted further from the game. It's an effect many former *EVE* leaders have talked about: the experience of no longer being part of your community but above it. "It changes you," one ex-alliance leader once said. Mactep's ex-associate said that by 2012 Mactep was rarely around anymore, and sometimes when he was around he was too drunk to be of any help.

"Mactep got used to treating people like shit – he has been compared to the British Queen, so no surprise that no one wants to talk to him," wrote Mactep's ex-associate BigMaman on the Russian *EVE* forums. "Especially to listen to his threats and stomping fits about who to accept or not to accept into the alliance. He would always tell Solar members what a cool diplomat he is and how smooth is the line of his politics, but they would never find out the real truth behind it all.

"Thing is, if some coalition at the right moment knew the real state of Drone Regions, they would have destroyed Solar Fleet and [Legion of xXDEATHXx.] Death wouldn't help Mactep and vice versa. But everyone believed in this best friends shit or were just frightened.

"One alliance leader betrayed the other and got money for it. This is the exact reason why two bears, living in the same place, stopped talking to each other. Completely."

DEATH AND THE TYCOON

The first time I spoke with Death he told me he had been afraid for his life. On a half-scrambled Skype call with one measly bar of reception, I strained to listen as Death explained a controversy which had scandalized the Russian *EVE* community in 2009 at the peak of the Great War. An ultrawealthy player had joined the game and started throwing real-world money around to reshape the nullsec political field.

"Evil Thug and Mactep got paid by the Russian tycoon," Death told me. "Evil Thug got $30,000 and Mactep got $15,000. Mactep got paid off, but the tycoon paid Mactep and Mactep promised him that I would stay away from the war too.

"But Mactep didn't share any money with me. He just took the money and told me I [should stop fighting in the war.]"

He explained his theory that Mactep had accepted $15,000 from "the tycoon" to withdraw from the war against Band of Brothers, and never told Death about the transaction.

"The tycoon was not a shy guy on the Russian forums," he said. "He did not hide that shit. He's one of those guys you know, instead of spending all his money on girls and in the clubs he wanted to show off in his nerd community. He was playing one of those browser games, and he ended up paying close to half a million [dollars] in just random

fucking pixel shit. He came to *EVE*, somebody brought him to *EVE*, and he wanted to play his grand strategy stuff. He ended up paying about, umm, he purchased some $300,000 in Titan characters. He paid Mactep plus this, plus that. He probably ended up somewhere half a mil. Plus or minus."

Death refused to follow the tycoon's commands because he never agreed to any deal and had no idea about the payment. This incensed the tycoon who thought he had been robbed.

"And I didn't know at that time that he paid Mactep," said Death. "Mactep only told me later on, being drunk, that he took the money, $15,000. But who knows how much he took in reality if that guy was so fucking mad. I assume he might've paid him much more, because he paid Evil Thug $30,000."

The scandal reached an abrupt halt, however, as the mysterious tycoon was apparently banned, and their purchased armada confiscated by CCP.

"The tycoon got all his characters and Titans banned," said Death. "So imagine, you've invested a half million into the game, and you lose all this shit in one day. So he thought that was my doing too. Because he thought, 'who else is gonna report me except the guy who took my money and didn't do what I told him.'"

The tycoon is a type of force in the *EVE* universe that is difficult to fully explain in this story. Not every powerful player actually chooses to make themselves known to the community. Some of them run banks, bot networks, or ISK casinos on 3rd party websites yet may not be well-known or tied to a single character name.

"[The tycoon] started to threaten me in real life," he continued. "He did not know that Mactep didn't give me any money. So he was very dirty mouthy. That kind of guy you don't fuck with, because he has a lot of money to kill you, yknow? It costs nothing. Killers been ordered a lot of times from Russia to fly to New York same day and kill someone for like $5,000. It wouldn't have been some sort of crazy from James Bond shit. It was very simple. So I filed a report with the FBI, and I explained the situation. Because I was scared. I'm telling you. I was scared."

Death alleges that he knows who it was that leaked the information to CCP about the tycoon's accounts: another high-ranking nullsec leader. But given that Death was worried for his life over these matters, it's probably best not to publish that accusation.

"Could you imagine I could get killed for this shit?" said Death. "I was really scared. I still am. Because that guy's got a lot of fucking money."

Death's full name in *EVE Online* is "UAxDEATH." The UA stands for "Underworld Assassins," his old Mechwarrior 3 clan from back in the day. It's unavoidably ironic that someone who named themselves, "Underworld Assassins, Death" ended up contacting the FBI for fear of being killed by an underworld assassin.

It's always difficult to know how much I can trust Death's version of events. His stories often adhere to the known facts which increases his credibility. Still, one thing he said to me always sticks out in my memory. When I asked him what he learned during his and Mactep's war to conquer the Drone Regions he said to me, "I learned how to manipulate people and what a great pleasure it is." He claims that the Russian community calls him "the king of the hypnotoads," explaining that hypnotoad is Russian *EVE* slang for a propaganda speech or essay that unifies members or twists a narrative, a reference to the classic meme. As with several of the alliance leaders I spoke to for this book, I occasionally have to wonder if it's wise to look into the eyes of the King of the Hypnotoads and take his stories literally.

Whatever the truth, when word leaked out to the rest of the Russian community that a rumored backroom payment had influenced Mactep, there was some public discontent, and according to Death, Mactep began to isolate Solar Fleet itself from the rest of the community.

"After Mactep told me about the $15,000, he couldn't shut the fuck up about it," Death said. "So he gathered up his second-in-command, third-in-command, his fleet commander and he told them. They were fucking pissed. They were enraged. He was being completely drunk, and I guess he couldn't hold his shit into him anymore. So he told that to his people and they quit. So people started to talk about him taking the money, and [Solar Fleet] kinda excluded themselves from the Russian community. Nobody excluded them, they just locked themselves out."

All of this was happening at a time when it seems obvious that a great deal of real money was flowing through the *EVE Online* grey economy, and Death characterized it as a common component of everyday life in the high-level gameplay of nullsec. Everybody was doing it, he says. Even he and Mactep.

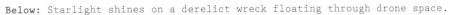

Below: Starlight shines on a derelict wreck floating through drone space.

"Mactep was selling ISK a little bit on the side, but he was never a big seller in those days. Just a little bit for himself, not a big deal."

Death noted that the ex-leader of Against ALL Authorities—Evil Thug, one of the most well-known characters in the Russian community—was infamous for making his living through *EVE*. Internet legend has it that former Goon leader Darius JOHNSON once accused Evil Thug of accepting money from the tycoon, and Evil Thug replied only, "New teeth, new flat, new life."

"Evil Thug was selling ISK for money, and he had never been shy about it," said Death. "He had never hid it from his members. People knew, he spends a lot of time in *EVE* and has to be compensated for it. Everyone was pretty solid about it, because he was honest about it. Everybody was doing it too. And it wasn't just the Russians. We picked it up from the Americans! How the fuck else would we know what the fuck is 'Ebay'?"

There was a culture of silence around the selling of in-game items or money, because it was considered bad for the in-game economy which harmed every player. It was also kept silent because it was a bannable offense that generally meant expulsion and ridicule.

"People would point at him and say 'oh look at the evil Russians they're doing it, they're doing it.' But meanwhile everybody else was doing it. Goons been doing the Ebay all the time. BoB was selling ISK. Everybody was selling it there was nothing wrong with it. If CCP was smarter they could have made huge amounts of money taxing it."

The culture of silence around ISK-selling was also fostered because the people who were doing it understood that just about anybody with a decent PC could do what they were doing and make a lot of money if they 1) understood that it could be done and 2) Googled how to do it. Nobody wanted anybody else to know they were making money because they didn't want competition driving down the market prices. Plus, the competition in the ISK-selling space was vicious. If an ISK-seller ever spoke publicly their competitors would certainly make sure that information made its way to CCP Games' security team. People only talked about their business when they quit years later and had no reason to remain silent. As time goes on, more and more people are opening up about the money they were earning. At the time, however, people rarely admitted it—it was an open secret hidden behind the thinnest veneer of plausible deniability.

"We had one Russian guy," Death told me. "His name was Loba. 'Evil,' in English. Or 'Pure Evil.' He had never been part of any group. It was just him and his brother. Very handy guys. He was making about $100,000 per month. This information came to me too late, because I would sell fucking ISK for real money too. I only created my alliance in late 2007-8 when a lot of money had been sold, and CCP cracked down real hard on people who were selling ISK. What I was doing was I was selling characters to Goons for real money. But that was nothing anywhere close to if I had started in those 2005-2006 times instead of being some stupid proud-of-himself idiot who was fighting non-stop and trying to be a fucking hero. I could've made a lot of fucking money. But with every year, every month a lot of people learned how to do it, and the profits went way down."

Throughout the late-2000s and early 2010s, CCP Games was engaged in proactive monitoring of external ISK-sellers, and searched for currency abusers in the data of the virtual universe. The biggest of these was the Unholy Rage operation which tracked and banned more than 6200 accounts suspected of being used to further RMT operations.

```
"Based on our new strategy and method-
ological process, CCP was finally pre-
pared for a major offensive against the
Real Money Trading (RMT) element in EVE
Online. Codenamed "Unholy Rage," the
operation was launched on June 22nd and
is still ongoing.
    In the weeks and months building up
to June 22nd we monitored and analyzed
activity of ISK sellers, and in partic-
ular, their supporting groups of mac-
ro-miners, ratters, mission farmers and
other RMT related units. The information
obtained by this research was then used
to identify further RMT type accounts
and to prepare lists to process in the
action.  The preparations also included
identifying various items of interest
that we wanted to monitor and measure,
such as market activity, server perfor-
mance [...] and so on, as our intent was
to examine the effects of the action in
```

It's possible the Tycoon got swept up in one of these operations and the story kind of faded away. Until one day in the Summer of Rage, one of Mactep's closest associates told the whole story.

THE BIGMAMAN HEIST

On July 22, 2011 a player named BigMaman—CEO of the executor corporation of Solar Fleet—defected. The turn came as a shock to much of *EVE*, who saw her as a loyal right hand of Mactep who was critical to the function of the alliance.

Her list of duties within the alliance was staggering. BigMaman once described her role in the alliance as "organizational and administrative work, management of the corporation, logistics, capital and supercapital building, espionage, [and starbase] management."

However, morale within Solar Fleet was strained after all of the rumors about the tycoon's $15,000, and according to BigMaman, Mactep wasn't taking it well. According to an account of the events on EVEOnline.com, BigMaman voiced concerns about the state of the alliance on their internal forums, concerns that she believed were being deliberately censored by Mactep. Her posting rights were revoked, and she was ignored.

"He has decided to show me his power and specify me my place," she told EVEOnline.com writer Svarthol.

BigMaman began planning to leave the alliance to get away from the deteriorating relationship with Mactep. However, she had joined the alliance in 2006, and had now given five years to her corporation. She believed fair compensation was in order. So on the way out the door BigMaman loaded up her wallet with 350 billion in raw ISK, and another 500-700 billion ISK in assorted goods, ships, and construction blueprints.

"I could do worse and clean [Solar Fleet's] bank, but I play honestly and never violated the EULA so I took maximum of what was legal," she wrote. "I could kick corps, disband both alliances and drop claims – but it will

have impact on others, people that I know and have good relationships [with]. But [my] problem is personal – and it's only about you. Besides you can easily forget to pay sov bills on your own."

BigMaman wrote an in-depth tell-all on the Russian forums after her defection.

```
"Lets start with the main topic - what is
the main reason for the conflict between
Mactep and UAxDEATH. It didn't start yes-
terday or before yesterday, it began 3 years
ago. [...] [Legion of xXDEATHXx] and SOLAR
FLEET were a formidable force back then,
and someone didn't like that very much.
The solution to this problem, to my sur-
prise, was as old as our world: money. And
the cost was reasonable for that man who
already had invested large sums into EVE,
$15,000. For the leader of the alliance,
by the side of whose I have been for the
past 5 years and considered him a friend,
and whom I have supported, it was enough.
    Considering that for the entire dura-
tion of our "friendship" he was unemployed
- myself and a few other people turned a
blind eye on everything. But in doing so,
we didn't see what would come next from
this. [...]
    During the war against Lotka Volterra,
The Five, Veritas Immortalis and others,
Mactep's presence boosted morale and won
such battles that there are still legends
circulating, but after those events, as
if sensing the smell of money, Solar Fleet
began to fail-cascade. [...]
    Since money has the tendency to end, and
since the virginity was lost, new methods
arose to fill up the pocket. All the fi-
nances were under control of one man, all
the decisions were made by him and he didn't
have to report to anyone either. First,
Rassatan's stolen Titan was sold for $4k.
[...] Then prices on supers dropped, but
there were now people who wanted to rent
space with real life money. I won't even
bother talking about RMT. When the system
```

works - who wants to give a chance for someone else to rule it? [...]

I was never ambitious and doubt I ever will be - I was still in Solar Fleet, because there were people who pulled the alliance alongside me. And of course, I considered him [Mactep] my friend. I was always understanding of his problems of unemployment, and only because of that I turned a blind eye and quietly did my duties. Fact is, I could disband both alliances [Solar Fleet and Solar Wing,] but also take all of the savings plus all the stations. I won't even mention banning all the shared accounts and those paid with bad credit card. I only took what I considered I had to. Yes, a few innocent bystanders got hurt, but couldn't happen without it. And I did leave enough in alliance wallet to make sure they could compensate those losses. If I was indeed paid to do this, SOLAR FLEET wouldn't exist for the past 6 months and Mactep would have died of alcoholism.

This is why I don't want to sit still and listen to all the brainwashing about me leaving being the reason for the war, it's just laughable."

— BigMaman, CEO, Solar Dragons, Solar Fleet defector
    July 23, 2011

With the vast riches of Solar Fleet in her pocket, BigMaman began to think about her future in *EVE* away from Mactep.

"When asked how she felt about the consequences for the corporation, she said 'Everything that I've taken away—will be easily restored, within 7-8 months or faster. I think—I have given a good lesson to the person how to correctly behave with people.'

BigMaman also mentioned that there were many people who had contacted her to show their respect and admiration and that she already had received offers from other alliances that would be interested in her work.

Solar Dragons [executor corporation of Solar Fleet] CEO Mactep refused to comment, saying the situation was 'banal' and that there would be 'no sense to speak about it.'"

— Svarthol, EVEOnline.com
    July 27, 2011

The total value of the take for BigMaman was estimated at the time to be worth about 900 billion ISK, about as much as one of the devastating brawls between the Northern Coalition and Drone Region Federation at O2O-2X. The grey market value of that at the time was $42,000. It was one of the biggest heists in *EVE Online* history, and just a month later, BigMaman took the gains and was invited to join Legion of xXDEATHXx. Mactep and Death were already at each other's throats, and it didn't help matters that Death was now openly harboring one of the most high-profile defectors of all-time.

All the fuel for a civil war was already there, it just needed a spark.

STAINWAGON

While this internal unrest was making waves in the Russian community, another problem developed for the Drone Region Federation.

The destruction of the Northern Coalition created a diaspora effect. The structure of the Northern Coalition had collapsed, but most of those corporations were still in *EVE*. Though it had defeated the Northern Coalition months ago, huge numbers of those Northern Coalition pilots and corporations were now trickling back to the game, having rebuilt their presence in *EVE* and found new allies. Many of those groups were assimilated into the Goonswarm-led ClusterFuck Coalition. But some remnants of Vuk Lau's former alliance Morsus Mihi had been chased from the north and stonewalled when they

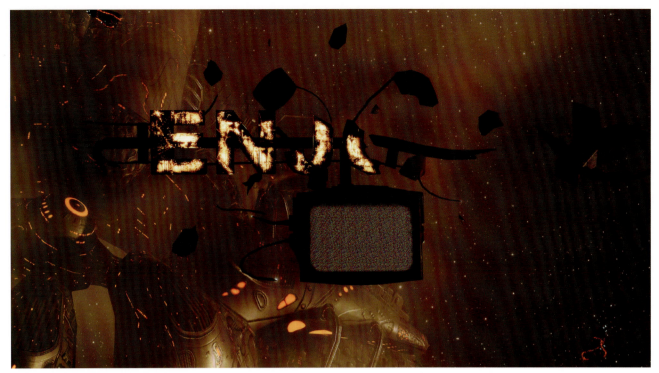

Above: A chunk of debris floats through Drone space. **Opposite:** The logo of the Stainwagon (RUS) coalition was a train for some reason.

attempted to challenge another group (Nulli Secunda) in the south. Morsus Mihi was on the brink of collapse, and decided to merge with Against ALL Authorities for the sake of the remaining pilots.

With AAA bolstered by an influx of several hundred experienced ex-Morsus Mihi pilots that shared a grudge with the DRF, the old alliance began to rumble back to life. Its famous leader Evil Thug had long since left the game, but a new alliance was being rebuilt under alliance leader Herculetz and fleet commander Makalu Zarya.

Following Morsus Mihi's lead, former remnants of Rebellion Alliance (who stayed together after it was disbanded by Daroh) moved to the South as well, seeing this resurgent Against ALL Authorities as a new vector for getting revenge on the Drone Region Federation. They were joined by two reclusive roleplay groups.

"Stain Empire and C0VEN, ancient groups that lived in and around the region of Stain for nearly a decade, emerged from hibernation," wrote Vik Reddy in his history of what would eventually become known as the Russian Civil War. "It was not the first time they lost their entire space. Because they retreated to the NPC region of Stain and lived in this corner of the galaxy since the game first began, they were incredibly resilient in the face of past invasions. Stain Empire and C0VEN were small in number and isolationist, steering clear of politics. They were elusive enough to be compared to the Ents from the Lord of the Rings series. However, it was an unusual time in the

game's history. They began to actively work with [Against ALL Authorities,] who they considered a friend, to combat the menacing DRF coalition."

Together this ragtag, largely Russian and European group of grizzled veterans and masterful roleplayers called themselves Stainwagon or the nickname "RUS," a nod to the Russian roots of its core alliance, Against ALL Authorities.

The coalition formed in order to oppose the massive and ever-growing influence of the Drone Region Federation, which now stretched across nearly half the territory in the game. Red Alliance in particular owned huge amounts of territory that used to belong to many members of the Stainwagon Coalition. Starting in November, the Stainwagon coalition went on a mission to take the south from Russian hands.

"This escalated the guerilla tactics to open up a second front that would focus on conquering Red Alliance's space, widely considered RUS holy land," wrote Vik Reddy. "The unexpected resurgence was dubbed the 'new Southern Coalition.'"

Tension gripped the southern regions as the Drone Region Federation was tested for the first time since the Northern Coalition war, and by a credible coalition of powers the DRF long thought defeated. The two met first at the Battle of LQ-AHE in Omist, the one-time home of the ill-fated Atlas Alliance and Ascendant Frontier.

"It was [Against ALL Authorities'] first major engagement against Drone Region Federation after the downfall of Morsus Mihi, an exodus that brought in several hundred new pilots' worth of firepower, including myself, under the command of Makalu Zarya," wrote Vik Reddy. "We were limited to a Drake battlecruiser fleet, a cheap but effective fleet doctrine at the time. The goal was to maximize damage at the least expense of ISK. [Against ALL Authorities] was still recovering from a long defensive campaign against the DRF."

However, it was the attack itself—rather than the result of the battle—which ended up being most important. Someone had finally tested the bonds of the DRF and the situation collapsed as they now took sides in a free-for-all for the South and the Drone Regions. Under considerable pressure, a spokesperson for Legion of xXDEATHXx announced that there would be a major diplomatic adjustment.

```
"There will be a standings reset be-
tween Legion of xXDEATHXx and Solar
Fleet right after downtime. This tough
decision came as a result of all diplo-
matic attempts, towards our old neigh-
bor Solar Fleet, failed. Drone Regions
are no longer the quiet corner of the
universe - a war begins now. Not only
the ingame war, but political and pro-
paganda warfare. Alliance Legion of
xXDEATHXx is now mobilizing forces to
fend off the aggressors and to protect
what Drone Regions stood for since the
beginning."

 — Kupyc, Legion of xXDEATHXx
```

What caused the Russian Civil War of 2012? For years afterward people would debate what had actually triggered it, but like everything in *EVE* it was a complex interaction that had been brewing for years. And to complicate matters we don't have all the facts.

The tycoon had created a rift between Solar Fleet and Legion of xXDEATHXx. BigMaman's departure turned the rift into a chasm. It was able to be covered up for months because nobody dared test the legendary friendship of Mactep and Death. But now the bluff had been called, and the two were forced to confront their animosity.

THE RESET

On December 8, 2011, Legion of xXDEATHXx and Solar Fleet wiped relations with each other, and a Russian civil war began. The first battle between the two occurred later that same day in GA58-7 in Outer Passage.

The first few battles established little else beyond the fact that both Legion of xXDEATHXx and Solar Fleet were ill-prepared to start an all-out war. It was soon obvious, however, that Solar Fleet was the vastly superior military group. A series of drastic Legion of xXDEATHXx battlefield losses ensued.

In the south, meanwhile, the Stainwagon coalition moved on Red Alliance and began a quest to retake its ancestral home.

Hoping to get the situation in the south under control, Death did what he often did when he was in trouble. He messaged Shadoo of Pandemic Legion, and offered the mercenary group a contract to come to his defense. Pandemic Legion was joined by fleet contingents from NCdot and TEST Alliance, the swiftly-growing alliance of Redditors from the ClusterFuck Coalition. The three alliances had been growing close recently, wreaking general mayhem across nullsec wherever they chose to attack. They really didn't give a shit who it was they were attacking or what risks they had to take. That's why they called themselves, "The Honey Badger Coalition." Did I mention it was 2012?

The ClusterFuck Coalition wasn't exactly pleased that its member TEST was fleeting with its blood rivals NCdot and Pandemic Legion, who had relatively recently attempted to sack the Goon capital. But CFC leadership reportedly believed that allowing TEST to explore other frontiers might strengthen their bond over the long term.

"Northern Coalition. and TEST Alliance helped Pandemic Legion, establishing the Honey Badgers," wrote Vik Reddy. "Its formation was meant to oppose [Against ALL Authorities'] efforts to gain a foothold into the heart of the Drone Regions. In their eyes, they were simply remnants of the old [Northern Coalition] who they were contracted to snuff out one last time."

CFC ATTACKS WHITE NOISE

No sooner had the Russian Civil War begun between Legion of xXDEATHXx and Solar Fleet than The Mittani announced that his DRF-allied neighbors White

Noise—who he had recently accused of rampant real money trading—were a dangerous threat and needed to be removed from the game for the good of *EVE*. On December 19, 2011 CFC fleets stormed the White Noise-held region of Branch under Fleet Commander DaBigRedBoat.

"In a Christmas Day meeting recorded on their Team-Speak comms, White Noise compared the situation to Germany's invasion of Russia in World War 2, citing similar mistakes due to lack of preparation and logistics issues," wrote Vik Reddy.

However, that was proven to be wishful thinking, and Goon fleet commanders Mister Vee, Lake, Lazarus Telraven, and in particular DaBigRedBoat, achieved devastating levels of success. Without the rest of the Drone Region Federation to back it up—occupied as they were with attacks from the "New Southern Coalition" and the civil war—White Noise was obliterated by a swarm of Goons larger and more uniform than ever before. In the climactic battle for control of White Noise's capital system, the Goon-led ClusterFuck Coalition fielded an awe-inspiring fleet of 1400 Maelstrom battleships that shocked the game.

White Noise was blindsided by the attack, because according to Death, The Mittani gave his word that it would never happen.

"The second that Solar Fleet went to war with me, Goons engaged my only trustworthy best friend White Noise," said Death. "So [when war broke out with Solar Fleet] those people could not help me even if they wanted to because they got stuck up, they lost all their assets locked up in stations, they got bubbled up and it was impossible to do anything. PsixoZZ got very mad, and he was fighting. He got very mad at me. I vouched. I said 'Mittani is never going to attack you, I'm giving you my word on it.'

"So Mittani gave me his word and he broke it," he continued. "Since that day we've never been friends again. We're political leaders. We talk, we chat, but we're not friends anymore."

However, much of the rest of the community saw it as a deft move, a seizing of an obvious opportunity to capitalize on the weakness of a geopolitical competitor. An *EVE* pundit named TheSpartan wrote at the time:

"In my opinion, the assault on (White Noise-held) Branch marks only the beginning of a campaign having as a main goal the establishing of the new sov dominant super-power in *EVE*. [Goonswarm Federation] correctly assessed the extreme weakness of their older opponent WN (White Noise). The supercapital nerf combined with White Noise's lack of leadership led to a very quick steamrolling of Branch by GSF and allies. We have to observe here that Goonswarm Federation were never friends with WN anyway. Therefore their action is a well calculated political and military move that at the moment establishes Goonswarm Federation as the major sov holder on the map. Mittani correctly assessed the situation and acted accordingly."

With the Drone Region Federation in utter turmoil, the future now looked increasingly like one that might be dominated by the ClusterFuck Coalition.

On January 9, 2012, The Mittani playfully mocked George W. Bush and announced "MISSION ACCOMPLISHED" in the White Noise campaign just three weeks after it had begun.

With White Noise taken out of the war by Goonswarm, Death was left alone to fend off Mactep's Solar Fleet. Pandemic Legion and its nascent Honey Badger Coalition were doing great work on the southern defense, but according to Death it was too little too late. Death credits the Honey Badger Coalition with working hard to protect his alliance, but says that his alliance was too weary after constant warring, and simple things became impossible. He took responsibility for the failure, and noted that any time Pandemic Legion needed any kind of administrative or logistical support to keep the war going, Legion of xXDEATHXx was having trouble making the simple stuff happen for them.

The Russian Civil War proceeded very poorly for Legion of xXDEATHXx even as Death continued to hire more and more mercenaries.

Death was looking to Red Alliance as his last ally. However, it was when Legion of xXDEATHXx needed its old friend most that the leader of Red Alliance—Silent Dodger, one of the most visible and well-known players in the Russian community—suddenly decided to leave *EVE*.

"I don't believe in coincidences," said Death.

SILENT DODGER CONFESSES TO RMT

"Somehow [The Mittani] found a way that Dodger could be paid off," Death alleged. "Dodger was doing eBay but he wasn't hiding it from people, just like Evil Thug.

So people knew about it. It was completely fine. But then Dodger got offered some solid numbers. I have no idea how much. I have no idea what he was offered, but he resigned as CEO like out of the fucking blue."

Before he signed off from the community for good, Silent Dodger wanted to be honest with the corp-mates he had spent years flying with. He wanted them to know the real reason he was leaving. In an all-alliance TeamSpeak meeting, Silent Dodger told his story.

DODGER: The main point, and I think that's what I'll begin with, is that I'm resigning as CEO of Red Alliance, and now I'll tell you why. [...]

I joined Red Alliance when no one was really in power, just after Fireknight stopped logging in and leading fleets and Razdalbay wanted to sell Red Alliance. This was when the alliance had no leadership, and I already loved it back then. I felt part of it. [...]

I then became the CEO, and in truth I became CEO because of Mactep. If you remember, we didn't have any allies back then and the situation with UAxDEATH wasn't clear. I then went to talk to Mactep, and he said, roughly speaking—"I won't deal with Fireknight, so why don't you become CEO". [...]

My vision for the alliance was the creation of an independent, influential and powerful alliance. [...] That vision became a reality towards the end of 2010. That's my personal opinion.

Ignoring all the drama, the rumours— some may have positive memories, some may have negative memories. What the alliance achieved in the political sphere, and the military sphere—I consider a respectable result. Looking back, I think I got the alliance to the level I wanted it to be at. [...]

I'll be honest—towards the end of the war against the Northern Coalition, I had absolutely nothing left to do in the game. When you become the CEO of your own alliance—your sense of self-importance goes through the roof. Whatever anyone says about that—it happens to everyone.

Eventually, PsixoZZ took all responsibilities onto himself. And when you give him your members, and declare that only one person will be in charge, and that is PsixoZZ—you lose all will to play, to command.

Red Alliance Member: Did you get upset?

DODGER: No, no, I didn't get upset. Take my word literally, I didn't get upset—I just lost all will to play, and began to stop logging in.

About the same time, my mother began to have health problems, and that is when I joined the ranks of RMT'ers—and sold my first Titan. It was the Ragnarok that you all remember.

Red Alliance Member: You do realize people are recording this?

DODGER: Yes, I know, I know. I don't care. Guys, I told you I will tell you everything as it is. I just want you to understand what motivated me.

Towards summer—I told you all I was going on vacation and would come back in a few months—and that didn't happen.

I went to work for the Samara local government. I lasted just 3 months. And that was when everything changed. It happened about the same time as the conflict with [Northern Coalition.]

Basically, real life finances became troubling due to debt—I bought a car on credit.

Obviously, I thought I would be making a lot of money—but when I went to work I understood that you have to work 20 years before you decide anything. It's a system where you are considered shit—and all your skills and knowledge don't matter. It's a job where you just have to obey—which my personality didn't allow me to do. This is why after 3 months I quit with my head held high.

Above: The avatar of Silent Dodger, leader of Red Alliance.

But I was still in a lot of debt. This is when I luckily, well "luckily," ran into a group of people who are professional [Real Money Traders.] They offered to work with me, and exchange ISK for real life cash. That's how it all began.

Over the course of about 3 months, I began to take cash from the corp wallet, my wallet, from gifts—whoever knew helped as they could. What happened then was that the ISK exchange rate collapsed—meaning more ISK for less money. Eventually in January I took 60 billion ISK, and I think if I hadn't done that things in the alliance would have been much better right now.

Concurrently, I began to notice that I couldn't lead fleets anymore. I didn't know what to do—what to tell people. My skill as a pilot fell to 0 because I hadn't played since the war against Northern Coalition. And obviously I couldn't demand anything of anyone.

Many of you remember my [Call to Arms] against SOLAR, when I welped (lost) a couple of fleets and stopped logging in. I was actually ill but after that I just

didn't want to come back. My exit now is because I still care about Red Alliance, but I understand that I've done bad and will do bad for the alliance. This is why I'm giving up my responsibilities, while things can still be fixed.

I won't say that I didn't have support—I had it difficult but no one knew.

I want the next CEO of Red Alliance, at least for now, to be Lenton, as I believe he's a competent person. His personality is very different to mine—he's much calmer and unemotional, but he'll make a great leader. So before your questions, I just want to wish you luck. I don't know what'll happen to me, but I'll leave my characters in the Alliance—if you let me.

I'll now focus on sorting out my problems in real life, which is why I won't be of any help to the alliance right now. That's it."

— **Silent Dodger**, Red Alliance February 8, 2012

The members of Red Alliance argued among themselves for a while after the speech about whether this was treason. One remarked: "Dodger I know who you are. I know how you led fleets. You got the alliance where you wanted, and respect to you for that my dear brother. But you're leaving the alliance during a war, on the brink of failure. And that seems like betrayal to me, no?"

Death says he knows why it happened.

"The war is striked and Dodger resigns as CEO and [appoints] two fucking guys I've never heard of," he said. "I can tell you their names. 'Ghost' and second guy is [...] Lenton Lust. Well, they came from Against ALL Authorities a few months before. So what those two did is they removed Red Alliance from Insmother and it became Against ALL Authorities' pet."

An Against ALL Authorities diplomat released a statement: "Dodger left Red Alliance confirmed. He made the announcement today. He also confirmed that he RMTed the alliance wallet for a while and [sold] his Titan. Red Alliance will not collapse. Red Alliance new leadership are fine dudes. They will see the changes through."

Above: The monument placed in C-J6MT by CCP Games to honor the system's rich history.
Opposite: Solar Fleet consolidates sovereignty in the Drone Regions and the southeast.

FINAL BATTLE FOR C-J6MT

At this point, Shadoo knew that the situation was likely irrecoverable, and he began preparing to give a grand send off to the war. He decided the final battle would be fought at C-J6MT, that most storied of server nodes. In the aftermath he filed a detailed report on the forums.

"I think everyone knows Red Alliance has decided to try a new start in *EVE* and re-locate — we wish them well and personally I hope they do well because *EVE* would be a poorer place without them in-game. Having already [evacuated] their corporate, personal, and supercapital assets from their once busy home system of C-J6MT [...] Red Alliance had no reason to defend the timer, and they had not asked for anyone else to do so either — nor had they called a [call-to-arms.]

So — we got together with Vince Draken (NCdot) to plan one FINAL HURRAH for this system so rich in *EVE*'s history. When I was but a noobie flying assault [frigates]

in the illustrious [Knights of the South] alliance — I had assaulted C-J6MT with Lotka Volterra/Veritas Immortalis/Chimera Pact/Knights of the South/etc — and lost. So it didn't go down well to leave the system [to] fall without an epic battle.

Luckily Vince agreed and we all decided tonight would be the unveiling of our new concept fleet — called Slowcats which Manfred Sideous himself had come up with. I was somewhat sceptical — I'm pretty sure so was Vince [Draken] but hey — fuck it, what's the worst that could happen? Unfortunately, [Against ALL Authorities] had read our call for capitals and were planning to call our bluff — by sending a batsignal to every [Russian player] out there that TONIGHT would indeed be epic. Everyone was calling [for players to set their] alarm clocks and I expected 600-700 dudes to face our ~150.

So I caught up with [Goonswarm Fleet Commander, Mister] Vee and asked him if he wants to 3rd party in a brawl of the month. Vee eager not to pass up a firefight of course obliged and began gathering a Drake fleet to brawl madly with. Unfortunately I had not realized it was a RUS holiday week apparently and the Against ALL Authorities/Red Overlord staging system alone had by now 500 dudes. And they had not yet even began forming. So Mittens types up a [Call-to-Arms] for more dudes with their version of C-J6 history and the reason we should all honor it with a ~Good Fight~ and suddenly 450 angry Goons began making their way in towards C-J6MT."

— Shadoo, Pandemic Legion

The Slowcat fleet was Pandemic Legion's latest fleet engineering invention which essentially made a fleet of carriers impossible to kill while also fielding thousands of cheap attack drones.

The defending forces of Red Alliance, the ClusterFuck Coalition, the Honey Badger Coalition, and their mercenary allies Ev0ke combined for 940 pilots. Plus, under Manfred Sideous' index finger were 2100 drones, enough to buckle any subcapital ship in an instant.

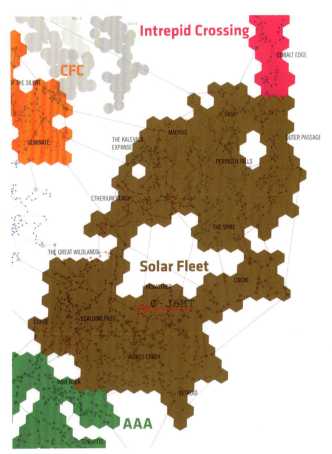

The attacking Stainwagon Coalition managed 1150 pilots in fleet for the symbolic final battle of the Russian Civil War.

"Solar Fleet greeted the 450-man Drake fleet at the YPW-M4 gate as they jumped in, brawling as the Drakes decloaked," wrote Vik Reddy who was personally present for the battle. "Mister Vee bloodied the Maelstroms and managed to hold the field, forcing the battleships to warp off and re-position. But he took substantial losses, especially to his Scimitar wing (support cruisers that help larger ships track fast-moving targets.) Meanwhile, the attacking 400+ man Tengu fleet from Against ALL Authorities engaged the Slowcat carriers while [orbiting] Makalu Zarya off the station undock. Despite raining down an extreme level of DPS on Manfred Sideous' Archon [carrier] we were unable to break his overheated tank as the rest of the carriers materialized and deployed remote repairs. Local [population] had hit over 1900."

Making use of its spy network, one Pandemic Legion tactic was to use a spy to discover who Stainwagon planned to focus its primary firepower on next, and then have that ship dock inside a friendly carrier ship, thus removing it from the field and making it untargetable. It was sometimes effective, but managing to maneuver in this amount of lag and time dilation proved tricky.

Both sides exchanged volleys of fire and tried as best as they could to maneuver in the lag and time dilation. In the end, a devastating bombing run by Fleet Commander Traderjohn of Macabre Votum (a faction fighting on the side of Stainwagon) turned the tide as it took a heavy chunk out of Mister Vee's Drake fleet. Eventually Mister Vee's ship as well was able to be targeted down. Lacking any more ability to effectively coordinate, the defenders reorganized and ceded the system to Stainwagon.

Shadoo took the time to write an in-depth battle report about the symbolic gudfite in C-J6MT.

"And as such, the EPIC fight draws to a close with ~600 ships in total killed to ensure this historic system does not fall without explosions," he wrote. "A BIG thank you for Team South, CFC, & all for keeping it interesting." Once in a while the strategic objective is just for everybody's pilots to have fun.

"Safe to say, now our problem is one fucking MASSIVELY SMUG [Manfred Sidious,]" he said, referring to the fact that despite losing the objective, the new Slowcat fleet concept worked perfectly.

CONCLUSION

"In 2012, Mactep took all of it," said Death, summarizing the lasting effects of the Russian civil war. "As you can imagine we were very tired after all of those wars. We were losing people left and right just because of exhaustion. You have to imagine that we've been fighting on average 13-16 hours a day. Not that many people can take it."

In March and April of 2012, Legion of xXDEATHXx was conquered by Solar Fleet, and Mactep obtained complete control over nearly all of the Drone Regions. Over the ensuing months, the Drone Regions were turned into a vast field sold by the constellation to renters on a weekly basis.

"[Legion of xXDEATHXx] retreated to low-sec, and they lost pretty much everything," said ProGodLegend of Nulli Secunda.

Red Alliance collapsed as well, and its new leaders led it to safety under the wing of Against ALL Authorities. Herculetz saw this as a golden deal. Red Alliance was no longer the threat it once was, and if it could be brought to heel it would make for a powerful shield to protect the Russian time zone from Solar Fleet's newly established empire in the Drone Regions. Herculetz offered Red Alliance the Delve region, so rich in history and income potential.

There was just one problem. Herculetz had already promised Delve to ProGodLegend of Nulli Secunda for helping them in the war against the DRF. Herculetz beseeched ProGodLegend to see the upside of this deal, but ProGod would have none of it.

With his hands tied, Herculetz offered a third option that nobody expected: Red Alliance and Nulli Secunda should fight for Delve. A sanctioned gudfite civil war.

DELVE 2012

What was conceived as a "friendly" civil war quickly turned into a frustrating mess as the wargame kept getting more and more serious between two coalitions who seemed to bring more and more pilots to every battle. When one side escalated, the other side had to counter-escalate, and a friendly shoving match turned into an all-out brawl.

That's when Pandemic Legion showed up and began shooting both sides. Montolio of TEST Alliance looked at his southern border and saw two potential powers fighting to become his new neighbor and his ally Shadoo of Pandemic Legion wreaking havoc on both of them. He announced TEST would be moving south to aid Pandemic Legion and annex Delve to house TEST's ever-growing membership. The young Nulli Secunda and its "N3" (Nulli Secunda, NCdot, and a smaller group called Nexus Fleet) coalition was largely overmatched by the massive number of TEST Alliance pilots, who piled ship after ship into Delve.

```
"One would do well to bear in mind that
anything with any sort of substance or
structure rising from Delve does not
bode well for TEST. With RED ALLIANCE's
expulsion from the Drone Regions, they
set about conquering Delve and Querious
all for themselves.

    Once RED ALLIANCE had set their eyes
on Delve, Nulli Secunda and other Delve
residents didn't take too kindly to them
barging their way into what appeared to
be a very civil tea party that had been
going on for months. This 'tea party'
bore no true ill will or threat to TEST,
but with the arrival of RED ALLIANCE
on the scene, for TEST to ignore this
omen would likely have been an error.
With boiling red clouds on the horizon,
TESTs silver lining likely arrived with
the news that Pandemic Legion would be
deploying to Delve to get involved in
the action, and so they have.

    Not long had The Legion landed on
the southern shores when they set about
descending from the rafters on NULLI and
RED ALLIANCE fleets alike. PL have given
literally no quarter and have taken no
side in the wrestling match between Nulli
Secunda and RED ALLIANCE for dominance
over Delve, and to TEST the arrival of
the Pandemic Legion killing machine is
a chance to hit several birds with one
stone."

    Ross Mcdermott, MMORPG.com
```

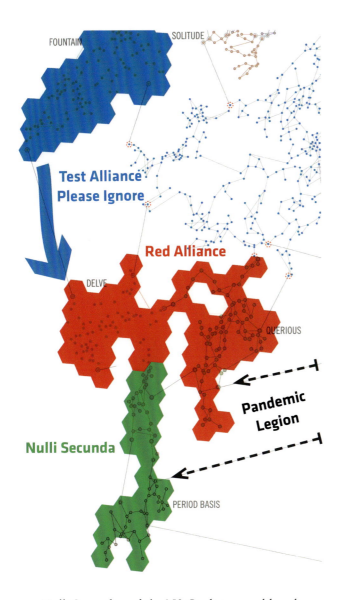

FOUNTAIN

SOLITUDE

Test Alliance Please Ignore

DELVE

Red Alliance

QUERIOUS

Pandemic Legion

Nulli Secunda

PERIOD BASIS

Nulli Secunda and the N3 Coalition could no longer hold off such a varied array of superpowers and pulled back from Delve to regroup.

In the aftermath, TEST led the ClusterFuck Coalition into conquered Delve and the resulting wave of enthusiasm propelled it into taking the nearby regions of Querious and Period Basis as well. As it did so, the CFC also stepped into the role of the dominant superpower in nullsec. The ClusterFuck's alliances now controlled all of the most profitable regions, by far the largest supercapital fleet, and a dominant economic position thanks to their technetium cartel and the ICE Interdictions. The Russian Civil War had provided the ClusterFuck Coalition with the leverage it needed to expand both its northern and southern borders, and in doing so it grew its already vast kingdom into the single vastest in the history of *EVE Online*.

But even in victory, the seeds of dissent were sown. Some members of TEST believed they had won the war single-handed and that the CFC had barely even helped. Montolio said he sympathized with that viewpoint, and so even though they'd just won a war together the relationship became surprisingly cold.

"The Glorious Goonswarm National Anthem"

With enthusiasm at a renewed high in Goonswarm, a Goon poet named Lyris Nairn sat down to finally pen a national anthem for the Goon state. Traditionally set to the tune of the national anthem of the Soviet Union, the song makes reference to many epochs of Goon history. In-jokes are rife throughout as the song references events like the time Goon fleet commander Mister Vee left his ship behind (a cheap Cruor-class frigate) inside a conquered station only to rally the alliance to go back and get it years later.

—

United forever in friendship and spaceships,
Our mighty Federation will ever endure.
The great Goonswarm nation will live through the ages.
The dream of its people in Fortress Deklein.
Long live our Deklein Motherland,
Built by Mittani's mighty hand!
Long live our New Bees, Rifters and Vee!
Strong in our friendship tried by fire.
Long may our Goonswarm flag inspire,
Waving with Fat Bee for space men to see.
Through days dark and stormy where Great Mittens led us
Our eyes saw the bright sun of conquest in Delve
and Mittens, Dear Leader, with faith in the People,
Inspired us to tear down the walls of Brick Squad.
We fought for the future, as conquering invaders,
and returned to our homeland the Cruor of Vee.
Our glory will live in the memory of spaceships
and all future nations will have ports that are free.
Long live our New Delve Motherland,
Built by Mittani's mighty hand!
Long live our New Bees, Rifters and Vee!
Strong in our friendship tried by fire;
Long may our Goonswarm flag inspire,
Waving with Fat Bee for space men to see.

Above: A player clone is ejected into space as its capsule is destroyed.

One night, Montolio opened up to Vile Rat during a meeting with the Goon diplomat. He and Montolio had met to discuss a number of issues including the construction of a station the two alliances planned to build together in the conquered capital of the Querious region, 49-U6U (the site of the Doomsday demonstration that marked the end of the Great War.) When Goonswarm asked for control of the station the TEST membership became inflamed. Goonswarm insisted it was an outpost at the front line of TEST space they could use to keep TEST safe, but to the TEST membership it was evidence of "imperial ambitions." In the following chatlog, Vile Rat's political skill was on full display as he lent a sympathetic ear to a stressed Montolio but also drew a hard line.

Montolio: going to have a dreddit directors meeting tomorrow, might be stepping down
vile_rat: Well that was unexpected
Montolio: my abrasive diplomacy may not
be what we need going forward
Montolio: [...] probably the last two
months of my dealings with you guys have
been super rocky and im afraid thats my

personal feelings getting in the way or
something
vile_rat: What do you think is causing
this rockiness
Montolio: I'm not entirely sure, cultural differences, difference perspectives,
experience levels, goals etc. who knows
but I feel like I've seriously damaged
the relationship and one of our core
values is our relationship with goons
[...]
vile_rat: I can tell you what test looks
like from my eyes if you'd like. perhaps
that will help with perspective
Montolio: sure
vile_rat: It is our perspective that you
guys desperately want to be out from
under our shadow both real and perceived
Montolio: we do, we don't want to be
under any shadow we are fiercely independent but also fiercely loyal
vile_rat: We don't consider it a shadow
and our efforts to be a friend to you are
met with a perceived iciness and arms

length We've never attempted to lord over test, we've never taken resources from you to keep for ourselves

Montolio: Right, you say "We are taking 49-[U6U] to protect you" But thats not what we hear

vile_rat: not to protect you per se but you asked the coalition to come down and help you take the region for really no benefit to the coalition at all fine. We're not even fighting that. but you have displayed a need for the coalition, it was our intention to put the coalition at the front line to discourage people getting uppity in your space Your guys took it as imperial ambitions and a desire to control you, and you as leader did nothing to discourage this thinking. It's become toxic and honestly it's become the reality in the eyes of the common test grunt

Montolio: Which is one of the reasons I am talking to my directors Because I sympathize with the test grunt viewpoint

vile_rat: And I contend it's based on a false premise and completely unfair to us and the work this entire coalition did to deliver you three regions on a platter with no expectation of any recompense

Montolio: Your motivations and goals are pure, but we'd rather be the ones attacked. If we can't defend these regions ourselves then its somewhat pointless

vile_rat: if our motivations and goals are pure, why do you help stoke the fire against us

vile_rat: you could say "no, they are just being friends trying to make sure we're safe in our new home

Montolio: I do say that

vile_rat: but you egg them on. encourage that kind of talk act like we're assholes for even suggesting such an arrangement

Montolio: But we can defend it ourselves, at the very least we'd like to

fail at defending it nobody has invaded us before [...]

vile_rat: what happens if you start to lose it.

Montolio: You guys lost several times and you seem better for it tbh. If TEST can't win in US [time zone] then we don't deserve it

As the conversation progressed, Montolio became more and more resigned and Vile Rat became more and more incensed at the lack of appreciation for all the CFC had done to get TEST to this point. But, Montolio explained, the relationship was destined to fail at some point. Goons and TEST were just too much alike, and now that TEST was fully grown, that was likely to mean independence. The unity of the CFC was hanging by a thread.

For now, three-fifths of the game rested beneath the rule of a rickety ClusterFuck led by The Mittani. For him it was a twisting, winding path to power that had taken seven years to achieve. The Mittani wrote to his members in celebration:

"We have not gotten nearly enough suffering from the conquest of Branch. We must have our screams; they'll just have to come from Empire. During the war, freelance Dread Pirates have continued to sow terror across Empire; the whispered implication of another Interdiction has kept the price of oxytopes at double their pre-Interdiction levels. We are not done with hi-sec, and our Finance Team (wizards of capital extraction, and clever as only sociopaths can be) have invented something that makes the ICE Interdiction look like amateur hour.

The Interdiction made the game shudder in horror. Our Victory Lap will make the Jita Riots look like a blip. [...] EVE will learn that when Goonswarm wins a war, the celebration will be the suffering of absolutely everyone else."

The Mittani, Goonswarm Federation, ClusterFuck Coalition
January 14, 2012 ●

BURN JITA

"We are going to go to the heart of high-sec, the beating
heart of *EVE Online* and we are going to stab it repeatedly."
— The Mittani, CEO, Goonswarm Federation

On April 27, 2012, in the Jita star system in the heart of empire space, Goon freighters were loading a ship hangar with 14,000 identical ships. Day after day, more and more of the identical ships arrived off the assembly line. They were heavily modified Thrasher hulls, rewired so that all the ship's energy focused on its weapon. No shield generators. No microwarpdrive. Just a Goon armada waiting in Jita—ostensibly the safest star system in *EVE Online*—with a single bullet in each ship's overclocked artillery gun. The ships were there to mount an all-out attack on the central hub of this virtual world.

Why? Because 30 days earlier, CCP Games had banned The Mittani from *EVE Online*.

FANFEST 2012

The story of the burning of Jita begins in Iceland on March 26, 2012 at *EVE Online*'s annual fan convention "Fanfest." Each year, the *EVE Online* community came together to meet each other in real life and give a series of presentations aimed at sharing knowledge and enhancing the common understanding of this spectacularly opaque and famously complicated video game.

It was also—in keeping with longstanding Icelandic tradition—an opportunity to consume spectacular amounts of alcohol. Throughout Reykjavik, the resounding cry of Fanfest weekend is "Skål!" the Icelandic "cheers."

Many of the game's most well-known figures were in attendance, and for three days in 2012 the community was united in celebration for a game at the peak of its

popularity. Those who had climbed and clawed their way to the upper echelons of EVE found themselves at the top tier of a game that was constantly ascending to new peaks. Few would have suspected when they began playing that they would one day end up speaking on-stage at a conference at the top of the world, covered by dozens of media outlets.

Of particular interest at Fanfest are the player presentations, in which some of the game's most well-known figures speak to an audience about what they've learned over the previous year about how to succeed in *EVE*. Famous fleet commanders present advice on fleet strategies, market geniuses speak about the secrets that made them digital trillionaires, and CCP developers present their plans for the future of the game.

In 2012, one of the centerpiece presentations was an alliance leader panel, in which many of *EVE Online*'s most popular player organizations were given a chance to introduce their alliance and talk about their group's story and ethos. Over the long 75 minute panel, six alliance leaders prominent in *EVE* traded stories and ill-advised shots of Jagermeister and Redbull. After an hour of talking and drinking, The Mittani's Goonswarm presentation was last. Within 24 hours of the presentation The Mittani would no longer be the chairman of the Council of Stellar Management, and he would be banned from *EVE Online*.

THE WIZARD'S HAT

The crux of his presentation—delivered while wearing a large purple wizard's hat emblazoned with gold stars—was based upon the half-sarcastic thesis that Goonswarm truly does love the rest of the *EVE* community. There's a lot of truth to this, of course, because the *EVE* community itself is Goonswarm's favorite toy, and yet everyone understood the sarcasm of calling what Goonswarm does "love."

During the speech, which lasted roughly ten minutes, The Mittani presented several case studies in which Goons had not only scammed, destroyed, or invaded other players, but relished their wailing cries in the process. Pretty basic stuff by Goon standards.

In a tone of deep mockery, The Mittani, (by now heavily intoxicated and continuing to slam Jager shots during the speech) read several of the communications he and his GIA agents had received from other players or intercepted from their forums. One was a post from a religious member of the old BoB ally Firmus Ixion who had posted a literal prayer to God to deliver the alliance from the hard times it had fallen on. He also read two notes that were from enraged players who had been taken by Goon scammers.

The most revealing anecdote was about a player from high-security space that some Goon bombers had discovered flying 22 mining ships all by himself during the ICE Interdiction. After the bombers destroyed the miner repeatedly, they then suckered the miner into purchasing a fake "protection program" then blew them up again anyway. The miner sent a message to the bomber crew at the peak of despair.

As The Mittani read the miner's impassioned plea aloud, thick with sarcasm and ridicule, the audience often chuckled along and the co-presenters occasionally egged him on, handing him more drinks. Yet others were aghast at what they were hearing. The miner's letter read:

```
"So now it looks like you will still
gank me, and I work hard to keep going
in this game. Sorry I am very mad [I
was going to sell those minerals to buy
subscription time] for my guys. Yes I
can make that back easily mining if I
could mine. Now I will just get popped
by you guys no matter what. [...] Now
I feel that I have been suckered into
giving away 1.3 billion ISK. Since my
divorce, all I want to do is die, and
I've been doing that a lot in this
game.
     I am sorry I did not understand. I
am just sick and tired of sitting here
alone and having to play with myself.
```

```
Everyone that I have helped out in this
game and in real life just takes what
you have and that's it. Never to hear
from them again. I am getting tired of
everything. It was nice mining ICE while
it lasted, took my mind off everything.
Even though some people may say I'm a
bot, I am not. I run all 22 accounts
myself. It is not easy, but it keeps
me sane.
     Sorry for making you mad at me. I
will leave you alone now and never enter
your space again. I will be off looking
for a nice quiet corner somewhere."
```

Many were shocked that The Mittani would mock a person who appeared to be in sincere distress, and yet this was not entirely unexpected from a delegate of Goonswarm. The culture of the organization has long been focused on exactly this type of griefing and scamming, and they partially exist in *EVE Online* specifically because the game and its developers don't technically prohibit those actions. Some of the audience was sincerely amused by The Mittani's anecdotes, because this was far from unusual at the time in *EVE* which was scantly moderated. In context with previous Fanfest player presentations which were occasionally spectacularly juvenile this wasn't completely unusual. In fact, Darius JOHNSON had given a pretty similar speech with a pretty similar blood alcohol content at Fanfest three years prior.

The Goon point of view would say that the humor in these anecdotes comes from the absurdity of a person allowing a game to become so important to them. The Goons of this era would likely have said they were doing this person a favor by helping them break their unhealthy attachment to spaceship pixels. It's part of an ideological battle that dates back to the dawn of online roleplaying games. There is a fundamental disagreement between players about how the simulation should be treated. Is this real life? Does any of this matter, and is it OK if it matters to you? Or is *EVE* a half-realm where everything is imaginary? Should virtual actions have a morality attached to them? Some players simply don't believe so, and believe that anything that occurs within the virtual realm is fair game. That's the very reason some of them are here. To exercise a demon without the risk it would carry in the real world.

There was a tenuous truce in the community where most conduct—even the sincerely vile—was permitted as long as it stayed within the virtual realm, and never crossed the line into real world harassment. Then that line was unambiguously crossed.

CRIME AND PUNISHMENT

At the end of The Mittani's presentation there was a Q&A session in which an audience member brought up The Mittani's story about the despairing miner. In his response to the question, The Mittani said, "Incidentally, if you want to make the guy kill himself, his [in-game] name is..."

While some in the audience and on the panel had laughed along with the entire performance others in the audience were furious, and within hours the story was picked up by gaming media who were covering the event. Overnight The Mittani's reputation experienced a drastic whiplash. No longer was he the devious emperor of space, the scribe of the well-read *EVE Online* column "Sins of a Solar Spymaster." Now he was, as one Kotaku.com headline called him, "the *EVE Online* Suicide Taunter." Cracked. com would eventually rank it among the most impressive dick moves in the history of online gaming.

CCP Games swiftly provided a statement condemning the speech and deflecting blame when approached by journalists about the controversy. However, CCP was on shaky historical ground. CCP had been advertising *EVE* for years as a cold, dark, and harsh space where anything can and did happen. Stories of players robbing, scamming, and griefing each other had been part of the marketing message of the game for as long as anyone could remember because those stories had a way of conveying the vividness of *EVE's* virtual frontier. When you encourage players to explore the boundary of virtual morality you shouldn't be surprised when one of them steps over it.

"I want to reassure you that CCP in no way condones the harassment of players, especially those who suffer from depression or suicidal thoughts, as we understand the possible consequences of such abhorrent behaviour.

Our Terms of Service (TOS) mirror our company's stance on this matter.

While the content of online interactions between players cannot realistically be gated within our game worlds, we do take very seriously accusations of such behaviour between our players.

Furthermore, we have a suicide hotline protocol which has, in specific cases, made a difference for several unfortunately troubled players. We appreciate you voicing your concerns on this level, and CCP will be very vigilant in monitoring any behaviour directed towards the individual named in the presentation.

We are undertaking a full internal review of this panel as well as the process used for vetting the panel's materials. Even though this panel was billed as unfiltered by CCP, we expect public presentations to be courteous and professional towards others."

The gravity of the situation clearly began to dawn on The Mittani as he flew home from Iceland to the United States. He issued a lengthy apology upon returning, and it opened up fascinating questions into where the virtual realm began and ended. Were The Mittani and Alex Gianturco two distinct people? If so, where was the line drawn?

"I feel absolutely ashamed of my behavior at the Alliance Panel. It's one thing to play a villain in an online roleplaying game—when I post on these forums or on Twitter, I usually do so as 'The Mittani,' and do my level best to convince everyone that I'm an unrepentant space villain, as that kind of facade provides an in-game advantage to me and my alliance. But I am not that character in real life, as anyone who has met me can attest. I went way, way, /way/ past the line on Thursday night by mocking the Mackinaw miner at a real-life event. I, as a person, am not the entity that I play in EVE; I am not actually a sociopath or a sadist, and I certainly don't want people to kill themselves in real life over an internet spaceship game, no matter what I may say or do within the game itself. CCP may say

'EVE is Real', but *EVE* is not real—and the line between the game and reality should not be overstepped.

I'm relieved to discover that the Mackinaw miner is doing fine and mining away, despite being blown up by Goonswarm in-game. He deserves, and he has, my heartfelt apologies—here in public as well as a private apology. There's no excuse for what I did—while some might try to use my inebriation as a mitigating factor, I put myself in that compromised mental state, and the guilt of that is entirely mine.

If I could go back in time and not have included the slide mentioning the miner, I would do so. While the *EVE Online* character 'The Mittani' would never apologize for any sort of villainy in-game, I myself, [...] feel utterly ashamed and sickened by my behavior."

— **The Mittani**, Goonswarm Federation

If there was one thing the community could largely agree upon, it's that the developers should not interfere with interactions between players unless they result in real-life threats or other very specific crimes. In this case, that hard line had actually been crossed, and CCP felt forced to take action. What action could realistically be taken, however? The issue hearkened back to something Raph Koster wrote about during the development of Ultima Online. "What sort of punishment is even possible for virtual crime?" he mused.

Did CCP Games have the power to ban the most powerful player in its community? What would happen if it tried? Couldn't The Mittani simply create a new character called "The Mittani." or "xXTheMittaniXx"? And how could CCP stop him from managing his alliance outside the game using services like Skype, Discord, Slack, Google Docs or any other services they have no control over? Did they even have the right to? The vast majority of that work took place outside of its virtual jurisdiction. By attempting to ban him, would they instead turn him into a martyr who could spark the next riot in Jita?

With all of this public attention CCP was forced to do something, and with few good options, it decided on what one might call a slap on the virtual wrist. CCP banned The Mittani from *EVE Online* for 30 days.

STATE OF THE GOONION

In the aftermath of the decision The Mittani called a State of the Goonion address. On March 29, 2012, more than 2000 Goonswarm members gathered on a Mumble server to hear a speech in which The Mittani explained what happened, what it meant for the alliance, and what would come next. It was a dramatic moment in which The Mittani's speech oscillated between a genuinely concilliatory tone and intense anger that popped his microphone audio. In the speech he attempts to move the Goons on from the controversy, and to detail the revenge plot masterminded by a Goonswarm leader named Aryth. It was a plot they had come to call "Burn Jita."

"I wanted to get everyone together here today because obviously I just got banned. I want to reiterate to everyone that this ban is completely meaningless to the function of this alliance and this coalition. [...]

I did something dumb at Fanfest. I did something really, really dumb. I apologized for that. That apology was real. I'm 33 years old. I have never slipped up like that before. [Lag interrupts speech] This is the first time in my life that I have had one of those regrettable 'I did something when I was drunk and I need to apologize to people' situations. The problem of course was that this happened on a live feed at a Fanfest that had a massive number of press from outside the game.

My apology was genuine. A lot of people in our coalition thought that was a troll, because it's very rare for me to break character and say 'I feel really, really terrible about this.' I don't normally do that kind of thing. I know it. Everybody knows it.

I fell on my sword, and I owned it and I did it proudly, and I'd do it again. [...] But I don't think I can be an alliance leader of Goonswarm effectively, and be the chairman of the CSM. This incident has shown that. I can either be a nice, upstanding citizen, and screw over Goonswarm by not being allowed to use the kind of tactics that are necessary to be able to have us succeed and survive in nullsec—which people in the media don't understand is like lawless Somalia, you can't show weakness in nullsec, you can't behave like a good citizen or you end up like Ascendant Frontier did, those wonderful citizens who of course got rolled over by a stronger power at the first opportunity.

So I resigned from the CSM. When I landed in Boston I felt terrible about what I had said and what I had done. I tweeted that I was going to resign and I didn't think that this was going to work anymore as part of my apology. And I did just that. [...]

Why did this happen? There have been many things said and done at many Fanfests before. This Fanfest was different from ones in the past, and I'm trying to interpret what has happened. It is true, I said a terrible thing, and I did apologize for it. This Fanfest had a whole host of media people who were brought here for Sony's major [Dust 514] launch. They weren't *EVE* reporters, they were FPS/Sony reporters. The entire crew shipped in more than 70 of them.

The intent that CCP has in working with Sony is clearly to grow the size of *EVE* and [Dust 514] by bringing millions of console players into the game world. This means that there was a lot of scrutiny at this Fanfest that simply wasn't there in the past as well as corporate interests that didn't exist previously.

Now, no, this isn't some sort of wacky conspiracy theory. It's hard to say if what we're seeing from CCP in reaction to what happened is coming from CCP themselves or from some other influence. We must wait and see—when I am unbanned—whether *EVE* is still the *EVE* we love. [...]

So here's what we're going to do. April 28th is the day that I will be unbanned. A month is a suitable amount of time to prepare, and to ensure that we have absolutely enough supplies and material to be able to burn Jita to the ground. And then we are going to do exactly that. We are going to annihilate Jita. We are going to go to the heart of high-sec, the beating heart of *EVE Online* and we are going to stab it repeatedly.

And if CCP is still the old CCP, this will be heralded as an amazing in-game event, a "Free Mittani" event, as it were, my first day out of the box. And everything will be fine. We will go on and we will continue enjoying the game—and it sucks that I was a terrible asshole and fucked everything up and I still feel really horribly guilty about it—but *EVE Online* will continue being *EVE Online*. If there is some sort of crackdown or reaction from Goonswarm being Goonswarm, then we will have more information and can make an informed decision as to how we proceed.

I know that after we have had conflicts in the past with CCP—with the T20 business, with the Threadnought—and in the past, my own reaction in 2007 had been extremely unreasonable. I started a raging threadnought against CCP because T20 was of course giving [Blueprint Originals] to Band of Brothers, [he] was in Band of Brothers, and we basically went to war with CCP. It was a disaster.

We're not going to do that this time. I have learned that rash action in this sort of context is *not* wise. It's not going to accomplish anything. You can't shoot CCP with a laser. And it might not even be necessarily their fault. It might not be the reasonable way to go. So what we are going to do is we are going to think, and we are going to keep this alliance and this coalition going, and then in a month, we will test the waters. We will destroy Jita. And we will see where things go.

For now, we have work to do. [...] All of today Raiden has come into our territory and [attacked more than 15 starbases.] [Inaudible] It is obvious to me that Raiden cannot be tolerated to exist on our northern border. If we are to exist as a coalition Tenal. Must. Be. Conquered.

I don't want you to unsub[scribe.] I don't want you to quit. I don't want you to rage impotently in Jita. Because the problem with this, if we do that, is our enemies will all benefit. They will see us having our downfall. They will point, and they will laugh. And at the end of the day we'll be letting our own people down.

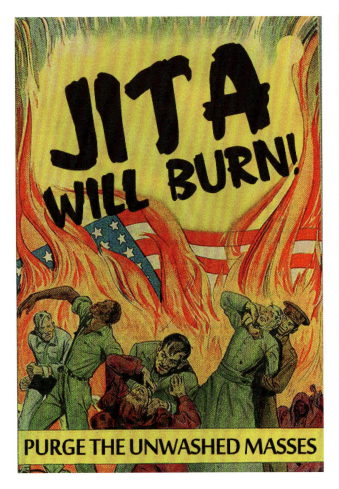

PURGE THE UNWASHED MASSES

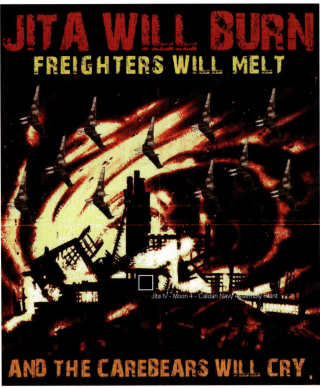

Jita IV - Moon 4 - Caldari Navy Assembly Plant

So I know. I know you're mad. I understand you want to point at CCP. I understand that I have been fucked. And I understand that it is shitty.

But we have work to do. And that work is in Tenal."

— **The Mittani**, Goonswarm Federation, Clusterfuck Coalition

March 27, 2012

It's a brilliant speech by any measure, delivered with dramatic timing and with tints of anger and defiance at all the right points.

The ban itself amounted to essentially a symbolic moment in which CCP needed to be seen taking action. It was not particularly meaningful to the bureaucracy of Goonswarm, roughly akin to banning the CEO from the company Slack for 30 days. It's not even particularly clear how CCP would go about banning a person who likely had multiple accounts, access to shared accounts in the alliance, and dozens of underlings who weren't banned and could relay information. But symbolic moments are often important because of the message

they send to the community. This speech was about The Mittani rewriting that message for his people.

The speech also deftly interweaves genuine *EVE* history with Mittani fiction. He uses the speech to reframe the narrative of the entire event. He explains that it is impossible to be the leader of Goonswarm and the chairman of the CSM, because he couldn't use the tactics needed for Goonswarm to survive in nullsec. But that was a smoke screen. This controversy had nothing to do with tactics in nullsec. In fact, he had recently boasted about what a success his first year on the CSM was, and how fruitful it had been "in the right hands." So he reframes the relationship as one that was destined to fail simply because he's a Goon.

Later, he references the T20 scandal of 2007, reminding his members that CCP is the real enemy who had been caught cheating four years in the past. However, he stretches the truth by conflating the T20 scandal with the "threadnought," which refers to something that happened five months later over a separate scandal in which CCP was largely cleared of wrongdoing. By fusing the two events together he creates a powerful new narrative: "In the past I was rash in the face of this kind of blatant corruption, but this time I will be wiser," he seems to imply.

The speech clearly had two primary driving goals. The first was to paint CCP as a company that had fundamentally changed due to its naked ambition and corrupt interests in trying to expand the *EVE* subscriber base. He raised the looming specter of CCP "bringing millions of console players into the game world." The other was to stabilize The Mittani's control over the alliance, and move people on from the event itself. Essentially, it aimed to bury the old controversy by making clear apologies, and then move the Goonswarm community on to a different controversy—one that had worked many times in the past: CCP malfeasance. After all, the entire *EVE* community was still on a hair trigger following the Summer of Rage in 2011.

A secondary goal of the speech was to give the Goonswarm community things to do and to look forward to which would carry them beyond the controversy. The task he gave them to occupy the moment was to retaliate against a resurgent enemy: Raiden, a bastion for old Band of Brothers directors. The Raiden campaign was never going to be a difficult campaign for Goonswarm.

Goonswarm took Tenal—historically speaking—in an instant. The sovereignty flipped in the Tenal region in a matter of days. The ease of this conquest makes Raiden's prominence in The Mittani's speech as an existential threat all the more ironic. When it was over, The Mittani installed RAZOR Alliance—one of the old cores of the Northern Coalition—back into their traditional home region. RAZOR had lived in the Tenal region for more than half a decade as part of the old Northern Coalition before the Drone Region Federation conquered its home. Now it was home again as part of the ClusterFuck Coalition. Much of the old Northern Coalition was back together under new leadership. Even while banned, The Mittani was advancing geopolitical goals.

THE SACKING OF JITA

While the Raiden campaign was underway, Goon logisticians and engineers built a massive stockpile of ships to facilitate the campaign against Jita. In order for the burning of Jita to work effectively, Goon pilots needed to know that their ships were disposable and free to replace. The whole thing would collapse if members were risking their own ships or were forced to fly all

the way back to nullsec to get a new one. And in order for their attacks to work and kill an enemy in a single shot, the ships needed to follow strict instructions in terms of how their armor, weapons, and capacitor power were allocated.

The burning of Jita was premised upon a loophole in the way the AI police force guards Jita and prevents it from being taken over by player groups. The CON-CORD police force warps in almost instantly to punish someone who commits acts of aggression against other players. Almost instantly. They don't stop the act from taking place, they respond to it. So Goonswarm's plan was to combine its two great loves—griefing and social engineering campaigns—to mount a massive ganking campaign indiscriminately targeting every single pilot who undocked from the newbie hub. Goonswarm built a massive communal stockpile of 14,000 disposable ships in the Caldari Navy Assembly Plant at Jita 4-4, and prepared to sacrifice every one of them in order to collect their most precious resource: the salty tears of "carebears" who play in high security space. It was part protest, part macabre Welcome Home party for The Mittani.

Thirty days after the ban was instated, in the early morning hours just after midnight on April 27, 2012, more than 1800 Goons made their way from their headquarters in Deklein and rendezvoused at the mountain of ships stockpiled in Jita for a fleet operation they'd been looking forward to for weeks.

The Goon fleet overwhelmingly consisted of Thrasher-class ships. Eighteen hundred of them floated in the sky above the Caldari Navy Assembly plant near the most heavily-trafficked trade hub in New Eden.

The Mittani had placed an enormous question mark over the entire Burn Jita operation: how will CCP react? In the past, CCP had celebrated player ingenuity on a grand scale. When players found clever ways to thrive and have fun within the rules of the game, CCP was well-known for celebrating that and letting players sort things out themselves. Their attitude in that regard is largely the reason why this book can be written.

Because this attack had been publicized by Goonswarm far in advance CCP Games actually had time to anticipate the event and shore up the servers. It even prepared to engage Time Dilation for the first time in high-security space in anticipation of an overwhelming event. Not only would CCP allow and endorse the deviously creative Goon offensive, it actively worked to facilitate it, and publicized Burn Jita as an example of *EVE* at its best.

In an interview during the event itself, CCP lead developers remarked that they loved the ingenuity of the players finding a unique way to spice up the game world without breaking the rules. These players, they noted, weren't evading justice. They were accepting it, and paying the accepted penalty: losing their ship. They'd simply come up with a plan so well-coordinated that losing their ship was almost meaningless.

The 1800 ships rendezvoused near the docking port of the station at the heart of *EVE*, and at the mark of the first fire command, 1800 Thrashers rained down a torrent of 1400mm Howitzer guns and smartbombs which obliterated a random freighter just outside the Caldari Navy Assembly Plant. Cargo and mining ships of innocent bystanders went up in smoke. CONCORD arrived and swiftly destroyed the ones who had fired, but within moments they were back in position in fresh ships and ready to do it all over again. Over and over again the ships were slammed with a hail of fire and then burst. The chaos continued for hours upon hours throughout the day, and kept up through the entire weekend amid a hurricane of Goon laughter, and a growing, glitchy graveyard of scrap metal picked clean by players harvesting the wrecks of ships destroyed in the calamity. Even the corpses of the dead traders killed were

often scooped up as mementos. The player would be resurrected in a new body, but these bodies would always be the ones that died the day that Jita burned. The Goon mission to bring nullsec culture to high security empire space was both a nightmare and a roaring success.

One anonymous player who was in Jita for the event summed it up better than anyone else. "There's just something special about building 14,000 spaceships and loading their guns with 1 round of ammo to shoot. And doing it right in front of the police," they wrote.

"This weekend has been a milestone for *EVE*," Lazarus Telraven, one of Goonswarm's top fleet commanders told EVEOnline.com. "The first time anyone has declared war on a solar system. The game worked wonderfully. Among the Jita Burners, complaints are few and praise is plentiful, but we all agree that without Time Dilation none of this would have been possible."

"It's what makes *EVE* a really good game," said Kristoffer Touborg at the time the lead game designer. "Do you want to play a 15 minute match of Call of Duty that you won't remember the next day, or do you want to spend four months manufacturing 14,000 Thrashers to do this? It's just so big and awesome."

The flames in Jita eventually dissipated, as the Jita Burners' numbers dwindled and they lost the critical

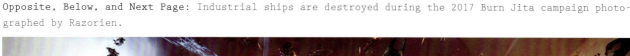

Opposite, Below, and Next Page: Industrial ships are destroyed during the 2017 Burn Jita campaign photographed by Razorien.

mass necessary to take out ships before being destroyed by CONCORD. The event was a hit, however, and has since become an annual staple on the Goon calendar.

For Goonswarm this was a landmark event, more important as a symbol than for the number of ships destroyed. For years now, The Mittani had been using large scale symbolic moves and backing them up with publicity campaigns to emphasize their impact. The ICE Interdiction, for example, wasn't important because of the miners destroyed which was actually quite low and ultimately irrelevant. What mattered was that the miners of high-sec believed they would be destroyed. Which stopped them from even attempting to mine. It was an intimidation campaign meant to augment the force projection of the spookily-named "Dread Pirates."

Burn Jita was essentially the same thing on a massive scale. It provided a clear statement that nowhere in *EVE Online* is untouchable, and that the very economy of the game was not beyond their reach. If the CFC could take over Jita one time then they could do it at any time of their choosing. Which effectively meant it was never truly safe.

This I think was ultimately Burn Jita's legacy. It raised a series of uncomfortable and fascinating questions about *EVE Online*. This time the question wasn't whether the

EVE community or CCP Games was in control of the off-switch of the game, as it had been during the Jita Riots, but whether a faction of players could gain control of it through force. And if not—as seemed to be the case, thankfully—could they harm quality of life within the game enough to affect subscriber rates?

In retrospect, this event has the flavor of an unspoken truce between Goonswarm and CCP Games.

When I look at this event I see two ambitions colliding. The first was the ambition of The Mittani who was clearly building the ClusterFuck Coalition into a superpower that could stand astride nullsec and atop the *EVE* community. He was peerless in his understanding of how to use forces outside of the game to affect the world within it. The Burn Jita campaign, in particular, was backed by a media push that again put Goonswarm and *EVE* center stage in the press. The event itself seemed conceived specifically to be attention-grabbing in the media so it could spread beyond the bounds of the virtual universe.

The second was the ambition of CCP Games, who dreamed of a digital *EVE* multiverse but needed to maintain *EVE's* massive cash flow to keep everything running.

The "Golden Goose" had to keep laying eggs every quarter or the whole thing would fall apart. Though they had heartily endorsed the Burn Jita campaign it raised the question: so what if they didn't? Could they have stopped it? Probably. But what would have happened if they had tried? They'd have turned The Mittani into a martyr, and risked exacerbating the Burning. And if it did begin to affect subscriber numbers then how should CCP explain such a situation to investors? "Sorry we lost money but the players are uprising?"

One of the things that separates The Mittani from other despotic characters throughout *EVE's* history is that he studied social engineering to better understand how to break the social bonds of his enemies. He understood that CCP Games might be devs but they were not gods. CCP was ultimately just another group of human beings. If you can understand how the group works then you can influence it. As a former Washington D.C. lawyer he intuited CCP Games' vulnerabilities as a corporation staffed by employees, backed by investor money, funded by a public audience, and he occasionally used that to apply pressure on the company and its employees.

Though it had been on a long journey over the past two and a half years since its collapse at the hands of Karttoon, the new Goonswarm Federation was looking stronger than ever, and the ClusterFuck Coalition was largely unchallenged. It seemed nothing could stop the rise of nullsec's new superpower. From a military and public relations perspective, The Mittani seemed invincible after recovering from a ban by spending three days as *EVE Online*'s first ever emperor of Jita.

However, humility would come soon, and from the most unlikely place imaginable.

Later in 2012, The Mittani will write to his members to tell them a story about where this journey in EVE truly began. He will write that he met up with his friend and alliance-mate Vile Rat—the infamous Goon spy and diplomat—in D.C. in 2006 while Vile Rat was in town to receive training for his real world job in the U.S. State Department. He will write that they went out to dinner, then drank vodka, logged into *EVE Online*, and flew their newbie frigates to 1V-LI2 to troll the old power Lotka Volterra. He will write that they concocted a scheme that night to infiltrate Lotka Volterra by convincing them Vile Rat had defected from Goonswarm and was looking for a new alliance. The Mittani will say that this unlikely sequence of events is what led Vile Rat to the battle where he witnessed the strength in Lotka Volterra's enemies: Red Alliance, as he watched from among the fleets that failed again and again to destroy the Russians at the Siege of C-J6MT. He will tell them Vile Rat was the man who sparked the six year adventure he and Goonswarm have been on ever since.

He will also write that he feels dead inside, and that he had made plans to see Vile Rat again before he left for Benghazi. ●

THE DEATH OF VILE RAT

"Sean leaves behind a loving wife, two young children and
scores of grieving family, friends and colleagues. And that's
just in this world, because in the virtual worlds Sean helped
create, he is also being mourned by countless competitors,
collaborators and gamers, who shared his passion."
— US Secretary of State Hillary Rodham Clinton

Because the story in this book often hops across the border between real and fiction in order to tell the sprawling story of *EVE*, it's worth stating explicitly that this chapter concerns a player's actual death.

On September 11, 2012, 34-year-old Sean Smith, the player behind Goonswarm's top diplomat Vile Rat, was killed in Benghazi, Libya in an attack that would become central to real-world events for years afterward.

The following chapter contains discussion of death and political violence. If you suspect it may cause you discomfort to read please feel free to continue on to the next chapter on pg. 181. The continuing story will not assume knowledge of this chapter.

SEAN

Sean Smith was born in 1978, and grew up in San Diego. He joined the Air Force in 1995 when he graduated high school, and became a ground radio maintenance specialist. He completed his military service in 2002 after achieving the rank of Staff Sergeant, and became a Foreign Service employee travelling to US embassies around the world ensuring that their computer systems worked properly.

"He loved computers," his mother told the San Diego Union-Tribune. "Computers were a part of him. You couldn't have one without the other."

Sean was also an avid gamer from a young age, coming up in online life in early text-only MUD roleplaying games where he had a special affinity for social manipulation. He reportedly met the woman who would later become his wife in an early online text-based game.

"Sean and Heather met in a MUD, one of the oldest of old-school multiplayer online games. It was an all-text game, but all Sean needed to make an impression was a few choice words. "The first thing he typed to me was 'You need to leave your guild or else I will kill you,' Heather recalled. "Classically diplomatic, even then."

Heather was a newbie player, just level five or so. Sean was angry at the guy who ran her guild. "So he was either killing all the people in the guy's guild or intimidating them," Heather said. "I was like, 'What do I care, it's not my fight?'" So she bailed on that guild. That was a classic Sean Smith victory.

"He was always trying to move the pieces and see how things went," Heather said. "He was really good at reading people to get them to either see his way or he could mediate in a way that he could get what he needed out of it—in a nice way, most of the time, I'm sure."
— From "The Amazing Life of Sean Smith, the Masterful *EVE* Gamer Slain in Libya," on Kotaku.com by Stephen Totilo

Opposite: A portrait of the avatar of Vile Rat, widely considered one of the greatest metagamers of all-time.

His job took him around the world to places like Montreal, The Hague, Praetoria, Baghdad from 2007-2008, and in 2012, Benghazi.

BENGHAZI

The circumstances of Vile Rat's death were a result of real world societal forces far beyond his control, and beyond the scope of this book's reporting. However, thanks to publicly available government reports we can better understand the circumstances that led up to it. At least, from the United States Government's point-of-view.

In early 2011, a year and a half before Vile Rat was killed, the Middle East was experiencing a series of protests, popular movements and uprisings across several countries. On January 14, the Tunisian government was overthrown by protestors, and on the 25th the world watched as demonstrators gathered in Tahrir Square in Cairo to demand the resignation of Egyptian President Hosni Mubarak.

Inspired by these events, the people of Libya also rose up against their own government. On February 15, 2011–while the Drone Region Federation was still mopping up Northern Coalition stragglers in the Battle of Uemon which had occurred just hours before—thousands of protestors gathered in Libya's second-largest city, Benghazi, to oppose the 42-year regime of Colonel Muammar Ghaddafi. When government security forces fired on that demonstration, it transformed into an all-out rebellion.

By March, the CIA had a liaison, J. Christopher Stevens, in Libya meeting with opposition leaders. Throughout the war, American counter-terrorist operatives were in Libya assisting the opposition with training in weaponry and tactics. Over the course of the next six months, the rebels gained the support of NATO and used air superiority to take control of the capital city of Tripoli and eventually the entire country.

The transition to the new government was chaotic. While Libya was nominally under the control of the revolutionaries, large portions of the country were said to be functionally controlled by a tangle of militias. The end of the war brought deeper United States involvement, as the State Department attempted to establish a permanent ambassadorship and gather intelligence on these local militias and their loyalties. The US built a temporary presence in the country on a reduced scale, while J. Christopher Stevens attempted to establish permanent roots for the US in a post-Ghaddafi Libya. The mission expanded to include a consulate building and additional staff, security forces, and an IT expert named Sean Smith, aka Vile Rat.

Meanwhile, the situation within Benghazi grew precarious.

```
"In the months [between February
2011 and September 11, 2012], there
was a large amount of evidence
gathered by the U.S. Intelligence
Community and from open sources
that Benghazi was increasingly dan-
gerous and unstable, and that a
significant attack against American
personnel there was becoming much
more likely. [...] The RSO [Re-
gional Security Officer] in Libya
compiled a list of 234 security
incidents in Libya between June
2011 and July 2012, 50 of which
took place in Benghazi."
    US Senate Committee on Homeland Se-
curity and Governmental Affairs Report
"Flashing Red: A Special Report on the
Terrorist Attack in Benghazi"
```

What we know for sure is that on the evening of September 11, 2012, Vile Rat was logged into Jabber talking with Goonswarm leadership and alliance mates, and he was already worried for his life. While chatting with friends he offered a caveat:

"Assuming we don't die tonight," he wrote. "We saw one of our 'police' that guard the compound taking pictures."

At 9:40 pm in Benghazi—7:40 pm in New Eden, and 2:40 pm in the United States where most of his friends were—two large groups of armed men approached the US consulate from different directions and began firing automatic weapons into the air.

We know the exact time because Sean was on Jabber, talking with his friends when they arrived.

```
[vile_rat 9/11/12 2:40 PM]: FUCK
[vile_rat 9/11/12 2:40 PM]: gunfire
```

A guard saw the groups approaching on a security camera, triggered the alarm and shouted "ATTACK, ATTACK" through the intercom system.

By this point Sean's alliance-mates were concerned, but those who knew him well knew this wasn't out of the ordinary for Vile Rat. While serving in Baghdad he would often be interrupted by gunshots and mortar attacks so they assumed he'd just come back in 5 minutes like nothing had happened. Like he always did before.

The militants threw grenades over the walls, and surged through the front gate by the dozens with automatic weapons and rocket-propelled grenades. Security agent Scott Strickland hurried Sean and J. Christopher Stevens away to a safe room in a different part of the compound.

Within twenty minutes the militia found the safe room, but couldn't find a way to break inside. With no way of cracking into the impenetrable vault, they torched furniture and lit diesel fuel fires around the exterior of the building. Blankets of dirty diesel smoke billowed into the safe room vents.

The three men in the safe room tried to escape the poison air through an emergency window, but only one of them, Scott Strickland, made it out. Strickland made repeated attempts to find Sean and the ambassador, but was unable to locate them in the thick smoke.

Sean Smith, known throughout New Eden as Vile Rat, died at about 10pm.

The US government scrambled to arrange a response. The only Americans in the world who knew about the attack in Benghazi were the highest ranking members of the US State Department, a few Navy SEAL Teams, and the Goons who happened to be hanging out on Jabber that night. The Mittani, one of Vile Rat's closest friends, was among them.

Though many now suspected something terrible had happened, it wasn't until about eight hours after the attack that the full truth emerged about what had happened.

"My people, I have grievous news," wrote The Mittani in the moment. "Vile Rat has been confirmed to be KIA in Benghazi; his family has been informed and the news is likely to break out on the wire services soon. Needless to say, we are in shock, have no words, and have nothing but sympathy for his family and children. I have known Vile Rat since 2006. He was one of the oldest of old-guard goons and one of the best and most effective diplomats this game has ever seen."

While news of the attack was trickling back to the United States in bits and pieces, within *EVE Online* there was already the beginning of an outpouring of grief over the death of someone the community now realized was not a wicked adversary, but a beloved leader in the community. Impromptu gatherings of hundreds of players on voice comms resulted in story after story pouring forth from grieving players offering condolences and sharing their favorite tales of New Eden's shadowy manipulator. Some noted simply and solemnly that they had made plans to share a drink with Vile at Fanfest next year. Others said they'd never met him, but were sad they never had the chance to know one beloved by so many.

Word soon reached American news networks, and the attack exploded into a major world crisis. Politicians cynically tried to use the event to spark a major scandal, as it had occurred on the eve of the 2012 US Presidential election. The denizens of *EVE* couldn't have cared less. They'd lost one of their own, and one of the rarest events of the digital age was taking place: a community-wide mourning.

By the next day, several nullsec factions had offered tributes, and a wave of alliances were beginning to rename their outposts in honor of Vile Rat. Goonswarm renamed the outpost in the capital of Tribute "We love you Vile Rat." TEST Alliance renamed the outpost at 49-U6U in Querious "RIP Vile Rat." The long-contested RA Prime in C-J6MT—presently under the control of Mactep's Solar Fleet—was renamed "RIP Vile Rat" as well.

Above: The character avatar of Vile Rat.

175

Over the course of the next day, a wave of grief poured over *EVE* as Vile Rat's story was shared around the world. The major alliances renamed dozens of stations as though they were candles left on the family's doorstep. Not for his nuclear family—his wife Heather and their two children—but for his space family: his close friend Alex, their thousands of alliance mates, and hundreds of thousands of fellow players.

Messages like "Farewell Vile Rat," and "We'll Miss You Vile" papered the star cluster. A number of stations said things like "Shoot blues > Tell Vile Rat" an homage to one of Vile Rat's legendary strengths: dealing with idiots who shot their own allies because they got bored.

These were beacons of respect which crossed factional bounds, because after nearly ten years in *EVE* loss was one thing that every alliance had in common. Most players I've spoken to who have been in the *EVE* community for a long time have been part of groups that lost someone.

The Mittani spent the night writing a proper eulogy for his fallen friend.

"So: Vile Rat, Sean Smith, my friend for over six years, both in real life and in internet spaceships, was the "State Department Official" killed in Benghazi […] Many were injured in these pointless, reprehensible acts, and one of my closest friends was killed as a result. […]

So. *Eve*.

[...] If you play this stupid game, you may not realize it, but you play in a galaxy created in large part by Vile Rat's talent as a diplomat. No one focused as relentlessly on using diplomacy as a strategic tool as VR. Mercenary Coalition flipped sides in the Great War in large part because of Vile Rat's influence, and if that hadn't happened GSF probably would have never taken out BoB. Jabberlon5? VR made it. You may not even know what Jabberlon5 is, but it's the smoke-filled Jabber room where every nullsec personage of note hangs out and makes deals. Goonswarm has succeeded over the years in large part because of VR's emphasis on diplomacy, to the point

of creating an entire section with a staff of 10+ called Corps Diplomatique, something no other alliance has. He had the vision and the understanding to see three steps ahead of everyone else—in the game, on the CSM, and when giving real-world advice.

Vile Rat was a spy for the Goonfleet Intelligence Agency. He infiltrated Lotka Volterra; he and I cooked up a scheme where we faked [Vile Rat] blowing up one of [a Goon's] haulers full of zydrine in Syndicate—this was back in 06 when zydrine [was expensive]—and that proved to Lotka Volterra that he had gone 'fuck goons'. BoB invaded Syndicate, then shortly thereafter GSF went to Insmother, allied with Red Alliance, and plowed over Lotka Volterra's territory, all with Vile Rat's aid. He came back in from the cold and became one of the most key players in the GSF directorate. His influence over the grand game and the affairs of Nullsec cannot be overstated. If you were an alliance leader of any consequence, you spoke to Vile Rat. You knew him. You may have been a friend or an enemy or a pawn in a greater game, but he touched every aspect of *EVE* in ways that 99% of the population will never understand. […]

Fuck. He was on Jabber when it happened, that's the most fucked up thing. [...] Then the major media began reporting on the consulate and embassy attacks in Libya and Egypt, and I freaked out and then it turned out that it was my friend of six years who helped build this alliance into what it is today, since the very beginning, starting out as one of my agents and growing to become the single most influential diplomat in the history of *EVE*, or perhaps of any online game.

I'm clearly in shock as I write this as everything is buzzing around my head funnily and I feel kind of dead inside.

I'm not sure if this is how I'm supposed to react to my friend being killed by a mob in a post-revolutionary Libya, but it's pretty awful and Sean was a great guy and he was a goddamned master at this game we all play. Even though a lot of people may not realize how significant an influence he had. It seems kind of trivial to praise a husband, father, and overall badass for his skills in an internet spaceship game, but that's how most of us know him, so there you go.

Shoot blues -> Tell Vile Rat.

RIP, my friend."

— The Mittani, September 12, 2012

As his remains arrived back in the United States, Sean was also eulogized by two other political leaders: US President Barack Obama and Secretary of State Hillary Clinton.

"Sean Smith, it seems, lived to serve, first in the Air Force, then, with you at the State Department. He knew the perils of this calling. ... And there, in Benghazi, far from home ... he laid down his life in service to us all. Today, Sean is home."

Secretary Clinton said in her remarks:

"Sean leaves behind a loving wife, Heather, two young children—Samantha and Nathan—and scores of grieving family, friends and colleagues."

"And that's just in this world," she added, "because in the virtual worlds Sean helped create, he is also being mourned by countless competitors, collaborators and gamers, who shared his passion."

It was remarkable to hear a sitting Secretary of State willing to nod to Sean's virtual life, but it was also a remarkable understatement. One could hardly blame Secretary Clinton for that. Encapsulating the virtual side of Sean named Vile Rat would take days and require a thousand storytellers to describe how his machinations shaped their lives in the subtlest ways. The images of the vigils that blanketed the stars of New Eden are the only adequate summary of the story of Vile Rat.

SEPTEMBER 12

In the same cruel way that the real world keeps turning after the loss of a loved one, so too did New Eden begin another day. Though it would do so without its beloved Rat, the wheel of *EVE* continued to turn, and Tranquility

Below: A meticulously created vigil left by members of the *EVE* community. Warp disruption bubbles have been anchored carefully above a station to spell out "RIP Vile Rat."

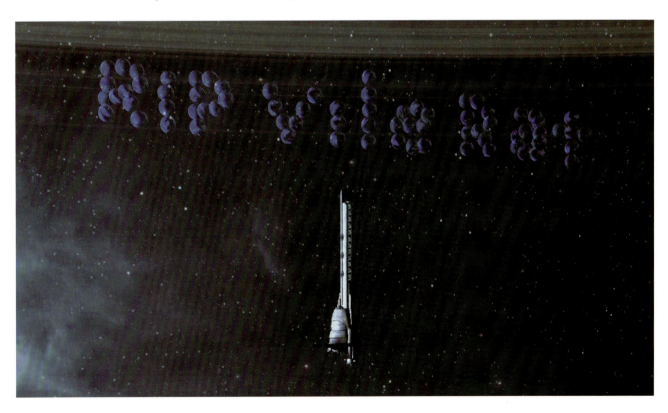

began a new cycle. There was a period of mourning out of respect, but the game had no choice but to progress.

New Eden eventually adjusted, and eased into a new norm. Over the ensuing months the CFC captured yet another region, Vale of the Silent, and expanded even further. But then the balance of power found a sort of equilibrium. The CFC was dominant while the N3 Coalition formed a sort of uneven counterbalance.

The major outlier was TEST and Pandemic Legion's Honey Badger Coalition. TEST was a member of both the CFC and the Honey Badger Coalition which allowed them to fly along with PL and get experience fighting—since dominant coalitions can often have trouble finding willing combatants to challenge them—and that was fine as long as those two loyalties never conflicted.

Until one evening in January 2013 when DaBigRed-Boat made a simple mistake that plunged the star cluster into chaos. ●

Next Page: A complete list of all the station names that were changed to honor the passing of Vile Rat. A common trope at the time was to name the system something that flowed from its random letter name. Many stations such as one in B-DBYQ would be renamed things like "B-Don't Forget Vile Rat."

-UUI5 IV - Moon 1 - Shoot Blues - Then Tell Vile Rat
H74-B0 III - Moon 4 - H74 Vile Rat Remembrance Station
3JN9-Q XII - Moon 3 - 3JNever Forget Vile Rat
60YQ-Z II - Moon 1 - RIP Vile Rat
C4C-Z4 VIII - Moon 3 - In Memory of Vile Rat
3BK-O7 VIII - Moon 1 - 3BK-07 Vile Rat Will Be Missed
BKG-Q2 VIII - Moon 1 - BKG0dspeed Vile Rat
5-6QW7 VII - Moon 3 - 5-6QWe Won't Forget Vile Rat
TPAR-G IX - Moon 3 - RIP Vile Rat
C-J6MT IV - Moon 1 - RIP Vile Rat
6-EVIQ III - Moon 1 - RIP Vile Rat
L-C307 VII - Moon 5 - RIP Vile Rat
9CG6-H VIII - Moon 4 - 9CGoodbye Vile Rat
K-8SQS VI - Moon 2 - o7 Vile Rat - May You Find Peace
NOL-M9 VI - Moon 2 - No Diplo Like Vile Rat
JU-OWQ VII - JU-st the Vile Rat Memorial
EC-P8R VII - EC-P8Rest in peace Vile Rat
P-2TTL I - In Memory of Vile Rat
FIO1-8 V - FA and EVE remember Vile Rat
J-LPX7 IX - Just Remember Vile Rat
QYZM-W V - Vile Rat Lest we Forget
LXWN-W IX - LXWNo Diplo could replace Vile
C3N-3S IV - C3Never Forget Vile Rat
Y-2ANO VII - RIP Vile Rat
7BX-6F V - Vile Rat Rememberance Station
OE-4HB V - RIP Vile Rat
PNQY-Y VII - PNQYou will be missed Vile Rat
G95F-H III - G95Forever Remember Vile Rat
YA0-XJ VII - YA0'll remember Vile Rat
ED-L9T I - ED-Love You Vile Rat
F-TE1T XV - F-Take Care Vile Rat RIP
I-7JR4 X - Thanks for the Memories Vile Rat
R-YWID VIII - RIP Vile Rat
T-M0FA IV - Test Mourns For Vile Rat
U-SOH2 VII - U-S0H2 Soon for VIle Rat
49-U6U IX - RIP Vile Rat
T-ZWA1 IX - DoP salutes Vile Rat
2-KF56 V - 2-K - RIP Vile Rat
15W-GC III - We love you Vile Rat
QY6-RK V - QY6-RIP Vile Rat
Y5C-YD VIII - You Will Be Missed Vile Rat
A1RR-M VI - RIP Vile Rat
Z-SR1I VI - RIP Vile Rat
I30-3A IX - Shoot Blues Vile Rat Memorial
7UTB-F V - 7UTBe At Peace Vile Rat
RG9-7U II - Everyone misses Vile Rat
8WA-Z6 VIII - We Will Miss Vile Rat
I7S-1S V - In honorem Vile Rat
C6Y-ZF III - C6Yes We Miss Vile Rat
0-HDC8 V - 0-HDC8 Shots For Vile Rat
PUIG-F V - PUIn Memory of Vile Rat
SVM-3K I - SVMany Will Miss Vile Rat
Y-OMTZ VII - Y-OU Will Be Missed Vile Rat
H6-CX8 III - RIP Vile Rat
GY5-26 IX - RIP Vile Rat
ME-4IU V - ME Gonna miss Vile Rat-RIP Mate
OWXT-5 IV - OWXTo you we raise a glass Vile
O-JPKH IV - LOVE YOU VILE
HM-XR2 V - HM-XRemember Vile Rat
QX-LIJ V - RIP Vile Rat
7T6P-C IX - Rest In Peace Vile Rat
DBRN-Z IX - DBRemembering Vile Rat
6F-H3W XI - 6F-H3We wont forget Vile Rat
ZOYW-O X - ZOYou will be missed Vile Rat
Q-02UL VII - RIP Vile Rat
EOY-BG III - EOY did it have to be Vile Rat
JC-YX8 VIII - Just Commemorating- Vile Rat RIP

Q-02UL VII - RIP Vile Rat
EOY-BG III - EOY did it have to be Vile Rat
JC-YX8 VIII - Just Commemorating- Vile Rat RIP
U-HYZN VI - U-HYZNo Replacement For Vile Rat
1-2J4P V - 1-2 The Memory of Vile Rat
PXF-RF V - Vile Rat Memorial
IG-ZAM VII - RIP Vile Rat
9DQW-W VII - Vile Tat Memorial
MZ1E-P IX - MZ1 Moment of Silence 4 Vile Rat
40-239 III - 4 Our Friend Vile Rat
CX8-6K IX - Can't Believe Vile Rat Is Gone
JI-LGM VIII - RIP Vile Rat
3L-Y9M V - RIP - Vile Rat
H-NPXW VIII - H-Never forget Vile Rat
Z-K495 XI - Hats Off To Vile Rat
2R-CRW X - 2R-Couldnt Save our Vile Rat
HB7R-F IV - RIP VILE RAT
W-IIYI X - W-IIYou Will Be Missed Vile Rat
T-IPZB VIII - In Memory of Vile Rat
0MV-4W VII - RIP Vile Rat we won't forget
L-6BE1 II - L-6BEveryone Misses Vile Rat
J1-KJP VI - RIP Vile Rat
HB-FSO VII - RIP Vile Rat
D2-HOS I - In Memory of Vile Rat
5-CQDA I - Farewell Vile Rat
CU9-T0 IV - CU On the other side Vile Rat
F-NXLQ III - F-NX Vile Rat Memorial Station
EL8-4Q II - EL8-4ever Rest In Peace Vile Rat
JP4-AA V - JPour one out for Vile Rat
I-E3TG V - I-E3To Our Friend Vile Rat
K5F-Z2 VIII - K5Forever remember Vile Rat
LIWW-P VI - LIll be missing Vile Rat
S-KSWL IX - RIP Vile Rat
FO8M-2 III - FO8M-2soon for Vile Rat
UMI-KK VII - U MIss Vile Rat
DBT-GB III - DBTo the memory of Vile Rat
30-D5G V - RIP Vile Rat
S-6HHN IV - Shoot 6 Blues For Vile Rat
BX2-ZX VIII - BX2-ZX Up If You Miss Vile Rat
M2-XFE V - Vile Rat - In Memoriam
FM-JK5 VI - FM-JKill Blues Tell Vile Rat
TEG-SD VII - TEG Vile Rat Memorial Station
HPS5-C II - In Memory of Vile Rat
C8-CHY X - In Memory of Vile Rat
3T7-M8 VII - 3T7 diplos do not equal Vile
WV-0R2 V - RIP Vile Rat
SY0W-2 III - RIP Vile Rat
F-88PJ VII - F-88 Ways We'll Miss Vile Rat
P-33KR II - In Memory of Vile Rat
Q-5211 VI - In Memory of Vile Rat
K4YZ-Y IV - In Memory of Vile Rat
B170-R VI - B170-Remembering Vile Rat
TN-T7T V - TN-This is for Vile Rat
RQH-MY III - In Memory of Vile Rat
B-DBYQ II - B-Don't forget Vile Rat
6GWE-A IV - 6GWE will remember you Vile Rat
BYXF-Q VII - BYXF-Quietly Mourning Vile Rat
MQ-NPY III - In Memory of Vile Rat
CCP-US XII - Thanks for the fish Vile Rat
O-PNSN IV - O-PNSNever Forget Vile Rat
BWI1-9 V - BWe miss you Vile Rat
1-5GBW VII - 1-5Go Rest In Peace Vile Rat
W-4NUU V - W-4NUwill be missed Vile Rat
W-IX39 IV - W-I Already Miss Vile Rat
UVHO-F II - UVHats Off To Vile Rat
CR-AQH III - Fujimo - Vile Rat RIP
LBGI-2 III - LBGI-2 Good to be gone Vile Rat

I-E3TG V - I-E3To Our Friend Vile Rat
K5F-Z2 VIII - K5Forever remember Vile Rat
LIWW-P VI - LIll be missing Vile Rat
S-KSWL IX - RIP Vile Rat
FO8M-2 III - FO8M-2soon for Vile Rat
UMI-KK VII - U MIss Vile Rat
DBT-GB III - DBTo the memory of Vile Rat
30-D5G V - RIP Vile Rat
S-6HHN IV - Shoot 6 Blues For Vile Rat
BX2-ZX VIII - BX2-ZX Up If You Miss Vile Rat
M2-XFE V - Vile Rat - In Memoriam
FM-JK5 VI - FM-JKill Blues Tell Vile Rat
TEG-SD VII - TEG Vile Rat Memorial Station
HPS5-C II - In Memory of Vile Rat
C8-CHY X - In Memory of Vile Rat
3T7-M8 VII - 3T7 diplos do not equal Vile
WV-0R2 V - RIP Vile Rat
SY0W-2 III - RIP Vile Rat
F-88PJ VII - F-88 Ways We'll Miss Vile Rat
P-33KR II - In Memory of Vile Rat
Q-5211 VI - In Memory of Vile Rat
K4YZ-Y IV - In Memory of Vile Rat
B170-R VI - B170-Remembering Vile Rat
TN-T7T V - TN-This is for Vile Rat
RQH-MY III - In Memory of Vile Rat
B-DBYQ II - B-Don't forget Vile Rat
6GWE-A IV - 6GWE will remember you Vile Rat
BYXF-Q VII - BYXF-Quietly Mourning Vile Rat
MQ-NPY III - In Memory of Vile Rat
CCP-US XII - Thanks for the fish Vile Rat
O-PNSN IV - O-PNSNever Forget Vile Rat
BWI1-9 V - BWe miss you Vile Rat
1-5GBW VII - 1-5Go Rest In Peace Vile Rat
W-4NUU V - W-4NUwill be missed Vile Rat
W-IX39 IV - W-I Already Miss Vile Rat
UVHO-F II - UVHats Off To Vile Rat
CR-AQH III - Fujimo - Vile Rat RIP
LBGI-2 III - LBGI-2 Good to be gone Vile Rat
38IA-E VI - 38IA-Everyone Misses Vile Rat
MA-VDX VII - MAny hearts to Vile Rats Family
5S-KXA XI - 5Shattered worlds RIP Vile Rat
O-BY0Y IV - In memory of Vile Rat
DW-T2I VI - Farewell Vile Rat - RIP
0P-F3K X - Rest In Peace Vile Rat
LEM-I1 V - VILE RAT RIP
O1Y-ED VII - In Memory of Vile Rat
KDV-DE III - In Memory of Vile Rat
2D-0SO XI - Rest In Peace Vile Rat
XD-TOV IV - Vile Rat Memorial
II-5O9 VIII - I wish Vile Rat was still here
JTAU-5 IV - LOVE AN MISS YOU VILE
F-NMX6 IV - In Memory of Vile Rat
RO90-H VIII - RO90oz poured out for Vile Rat
R3W-XU VIII - RIP Vile Rat
C-C99Z III - C-Can't Believe Vile Rat is gone
4-GJT1 IX - 4-Goodbye Vile Rat
UQ9-3C I - We Won't Forget - Vile Rat
3KNA-N II - S2N Salutes Vile Rat RIP
QPO-WI III - QPO-WIll never forget Vile Rat
43B-O1 VIII - 43Be well our friend Vile Rat
J7YR-1 V - Vile Rat Diplomacy Center
0B-VOJ X - RIP Vile Rat
V-LDEJ III - RIP Vile Rat
UI-8ZE X - In Memory of Vile Rat
60M-TG I - 60Ways Vile Rat Changed EVE
K-6K16 V - K-6 Vile Rat Remembrance Station

ASAKAI

"Sala Cameron and MrBlue had been working on [surprising] a small CFC
Sub-Capital group lead by Dabigredboat for about 3 days now, and by
looks of it tonight was the night when all the stars aligned. With
nothing at stake both sides threw caution to the wind and went all-in."
— Shadoo of Pandemic Legion speaking to Alizabeth for TheMittani.com

Many of *EVE Online*'s biggest battles and most important turning points were the result of years of planning and diplomacy and tactical maneuvering. This one happened out of the blue and for what seemed like no good reason in the middle of nowhere. The ships that are destroyed won't have a drastic impact on the fleets of any side, and the result won't markedly change anything. But the psychological impact it will have and the circumstances that will result from it will be profound. It began simply.

On January 25th, 2013, a fleet made up of two small allied alliances called Drunk N Disorderly and Lost Obsession was flying around space looking for something to do. The fleet meandered from system to system throughout low-security space, looking for a suitable fight. Not much was going on in this part of New Eden, and the three dozen or so pilots were searching for anything fun to occupy the evening.

After hours of searching for a suitable fight for their combined forces, a scout relayed word that ClusterFuck Coalition fleet commander DaBigRedBoat had been spotted nearby leading a sub-capital fleet.

Drunk N Disorderly and Lost Obsession didn't like Boat, and decided to screw with him a little. Their much smaller fleet made a detour to the system Boat was staging out of. On a lark, the Drunk N Disorderly gang surprise-attacked Boat's much stronger force, and inflicted

a bit of damage before Boat called up his emergency reserve of carriers to push back the attack. Little more than light-hearted harassment. The Drunk N Disorderly/ Lost Obsession fleet took heavy losses, but the cheap ships wouldn't cost that much to replace, and it was worth it just to mess with DaBigRedBoat and make something, anything exciting happen on an otherwise dreary Friday night roam through lowsec. The pilots laughed among themselves imagining Boat panicking and getting all mad. They flew back to their alliance's HQ station and logged off for the night.

But DBRB was less than amused that he'd had to call in carriers for a silly slap fight, and he decided he needed to make a point. So for days afterward he kept one eye open, searching for Drunk N Disorderly so he could send a message as old as power: you can't mess with us and get away with it.

Unbeknownst to either side, however, the political situation was about to get more complicated than anyone had anticipated.

While Boat was searching the nearby systems for Drunk N Disorderly fleets to make an example of, he received a message from one of the group's local enemies: the Liandri Covenant. Liandri ambassadors said that their alliance would be willing to keep tabs on Drunk N Disorderly's movements if DBRB agreed to have his pilots on standby to help out if Liandri itself was attacked as a result.

Drunk N Disorderly soon figured out Liandri was spying on them, and predictably began planning a revenge attack to destroy Liandri's staging base in this region. But since Liandri now had the backing of the CFC, an attack was exactly what they were hoping for. Liandri Covenant would just call DaBigRedBoat, who would reinforce them with a fleet of carriers.

Drunk N Disorderly knew it needed an ace-in-the-hole. So its leaders quietly contacted Pandemic Legion and told active fleet commander Hedliner that if he made his supercapital fleet available at the time of the battle there was a chance that he would get to surprise an ill-prepared CFC carrier fleet. Drunk N Disorderly said it'd have a Heavy Interdictor ship (used in preventing enemy ships from escaping) on the field to trap anything the CFC brought that was worth killing. What DND didn't mention was that because this was a spur-of-the-moment operation it only actually had one Heavy Interdictor.

To review: Drunk N Disorderly pissed off DaBigRedBoat for funsies. Their rival Liandri Covenant took advantage of that by offering to spy on DND's whereabouts so DaBigRedBoat could get his revenge. DND learned Liandri and the CFC were trying to set a trap for them, and saw an opportunity to lure DaBigRedBoat into a counter-trap. Drunk N Disorderly contacted a nearby Pandemic Legion fleet to be on standby in case DaBigRedBoat tried to intervene with his carrier fleet.

Liandri Covenant mentioned— maybe bragged—to one of its small allies that big things might be happening tonight, but that comment leaked the entire operation. A spy overheard the details and relayed that information back to DaBigRedBoat, who now knew everything Pandemic Legion was planning. In other words, everyone knew that everyone else had set a trap for everyone. And that made them all that much more interested in triggering it, because everyone thought their trap was the better one.

DaBigRedBoat requested a supercapital force to be on stand-by in case Pandemic Legion showed up.

BLACK RISE

On January 27th, 2013, a fleet of Drunk N Disorderly players was forming up and preparing to siege a Liandri Covenant starbase orbiting the 14th moon of the 4th planet in the Asakai system, in the Kurala constellation of the region Black Rise.

The starbase itself was mostly worthless. It had been placed there to serve as a forward operating base for Liandri Covenant in *EVE's* "Faction Warfare" territory (a more structured wargame zone that sees players fighting for the in-game lore empires.) Even for Faction Warfare space it served little strategic purpose or value. Tiberius Stargazer, a member of Liandri Covenant wrote, "It couldn't have even called itself an ammunition cache, because it didn't have any, it didn't even have any guns."

It was this empty tin can of a starbase—operated by an alliance that had little real influence—that would be the catalyst for one of the biggest battles in *EVE* history.

DABIGYELLOWTITAN

A setup attack the previous day by Drunk n Disorderly had put the worthless starbase into its reinforced mode. It was set to come out of its reinforcement cycle and become vulnerable at about 1am GMT.

"The POS (Player-Owned Starbase) came out of reinforcement and the Liandri Covenant fleet of some 30 cruisers and frigates waited for their enemies to show, and they did," wrote Liandri's Tiberius Stargazer. "Drunk N Disorderly, according to released battle reports had, due to an administrative oversight, almost forgot about the [battle and] had only engaged due to sheer boredom. DND, knowing Liandri fleet tactics, fielded smart-bombing battleships which countered a portion of the Liandri fleet's frigates leaving only the remaining cruisers to duke it out with the DND battleships."

As the small battle unfolded, DaBigRedBoat kept his fleet in a holding pattern several systems away, waiting for the right moment to open up a cynosural warp bridge into Asakai, bridge in his carrier fleet, and exact his revenge on Drunk n Disorderly and Lost Obsession. Keeping idle fleets entertained and attentive to fleet communications was said to have been DaBigRedBoat's legendary strength. He'd tell stories, crack bad jokes, and screen movies that the fleet would sync up and watch together until something happened.

To lead the fleet he was using two *EVE Online* characters at the same time. One account ("DaBigRedBoat")

LIGHT PROJECTILE WEAPONS

DESTROYER // MINMATAR REPUBLIC
INTERDICTION SPHERE LAUNCHER
SABRE

WARP INTERDICTION MODULE

Opposite: A Titan absorbs a barrage from all directions as forces continued to swell and both coalitions called for allies, or what *EVE* players at the time would have called "batphoning." Opposite: Sabre-class interdictors were commonly used to stop enemies warping away.

was in the carrier fleet that was waiting to warp into the system, and the other account was his Titan character Oleena Natiras. Oleena was logged in just to create the portal through which DaBigRedBoat and his carriers would bridge into Asakai where the Drunk N Disorderly fleet was just beginning to take control of the battlefield around the worthless starbase.

"After a couple of running skirmishes it became clear Liandri [...] did not have the heavy guns to break the formidable tanks of the larger ships," wrote Stargazer. "Out-gunned, the first call was made and was answered by the CFC."

When the Liandri Covenant fleet began to disengage, DaBigRedBoat's moment came. He looked at his screen and hit the "jump" command to warp his carrier fleet into the battle.

But there was a problem: he was using two monitors and clicked on the wrong screen. Rather than jumping the carrier fleet, he jumped his Titan—far more expensive and far less prepared for battle—leaving the carrier fleet behind. The only ship that arrived near the starbase was the transport Titan flown by Oleena Natiras, a 2.3 billion kilo surprise for the combatants in Asakai.

The pilots from the two low-sec alliances stared in amazement at the sheer size of the Titan which they assumed was just the first of many CFC Titans about to arrive. Titans usually only travel in convoys.

"A number of Liandri's rookie pilots (myself included) chattered in awe," wrote the Liandri Covenant pilot Tiberius Stargazer. "I hadn't—many had never—seen anything larger than a carrier."

Drunk N Disorderly quickly sent word to the on-call Pandemic Legion fleet commander, Elise Randolph, that the intelligence was accurate, but that something far juicier had appeared on the grid: DaBigRedBoat's Titan, alone. Within moments, the Drunk N Disorderly Heavy Interdictor burned toward the Titan, activated its warp scrambler, and trapped DaBigRedBoat in place.

"Boat's [original] plan was simple: hot drop the militia forces in Asakai, kill what he could, and get out," wrote *EVE* writer Alizabeth on TheMittani.com. "Once on field, he was tackled by DnD [Heavy Interdictor], who reported the Leviathan to Elise Randolph of Pandemic Legion. Elise decided to go for broke and committed 40 Pandemic Legion [supercapitals] to the fight."

DaBigRedBoat saw he was pinned down by the Drunk N Disorderly Heavy Interdictor and panicked; This ship was equipped for bridging carrier fleets, not for surviving bombardment.

"ALL CAPITALS SUPERS TitanS EVERYTHING LOGIN JUMP TO MJI3 THEN JUMP TO FIGHT SIEGE GREEN 20K FUEL," he posted in a flurry, ordering his entire fleet to come to his aid and attempt to save his Titan.

Above: A screenshot from the Battle of Asakai as the frantic melee reached its peak.

They were able to arrive quickly because Asakai was essentially on the doorstep of the CFC. But it was a fairly long trip for Pandemic Legion and those of their southern allies. So the CFC reinforcements (including DaBigRedBoat's carrier fleet) arrived in system just ahead of the first wave of Pandemic Legion supercapitals.

"Once PL supers were committed, jabber pings went out all over the CFC for fleets. The operation had turned from saving Boat's Titan to destroying the PL fleet. Subcapital fleets were formed and capital pilots logged in. One of the effects of Time Dilation (TiDi) is that it affects not just the system the fight is taking place [in], but several systems out. Capital ships were able to arrive much quicker than usual since they jumped in from outside of TiDi effects. When PL saw that the CFC was going 'all in' to Asakai, they reached out to other groups. They started out by contacting the N3 Coalition. When N3 agreed to assist, PL contacted TEST who started to make their way up north as well.

However, neither group was initially close and Asakai is located in the CFC's front yard, so to speak. It was a simple matter for CFC capitals to use cyno beacons to move to the battle and CFC subcapitals could use the highly developed jump bridge network. So, in the initial stages, the CFC had local superiority and was able to down one PL Nyx while three others warped out in low armor and even one at eighty seven percent structure. [The system population] was well over a thousand and climbing."

— **Alizabeth**, writing for TheMittani. com

When the TEST Alliance fleet left its base in 6VDT-H in Fountain with 500 pilots headed to Asakai to kill DaBigRedBoat's Titan because it was hilarious, Montolio reportedly said only: "Make me proud."

DaBigRedBoat's allies streamed into the system to help extract his Titan, but the arriving pilots were terrified to find a growing Pandemic Legion fleet of more than 40 Titans and Supercarriers. CFC fleet commanders were undaunted, however. They ordered their pilots to form up fleets, sound the "Horn of Goondor," and get everyone in a ship, because something big is going down tonight in Asakai.

A great stirring was happening across New Eden as word started to get out to the rest of the playerbase about the bizarre spontaneous battle that had broken out in low security space just a few jumps away from the newbie zones. With Time Dilation now in full-swing, even light-speed laser blasts streaked slowly across the tangled mess of the emerging battle, and missile barrages crawled in slow motion toward their targets. All sorts of pilots were intrigued, and a lot of them got the clever idea to journey out to the battle for themselves to take a peek, just to be a part of the commotion between great nullsec powers. This was the type of battle they'd read about in the news and attracted them to *EVE* in the first place. Now that it was so nearby, and with time moving at such a pace, people would have hours to get online, travel to Asakai, and see it for themselves.

It quickly began to dawn on both the CFC and Pandemic Legion fleet commanders that this was turning into something big, and both sides could now plainly see that with time itself slowed down, the only strategy for winning this impromptu battle was going to be to out-escalate the other.

The race was on between the two juggernauts to get their people organized and into Asakai before the battle reached a tipping point or the server crashed. Pandemic Legion leadership knew they had a strong chance to win the escalation battle because it was offering the head of DaBigRedBoat as a symbolic prize. As a long-time Goon fleet commander, Boat was someone a lot players all over New Eden wanted to see humbled. As an added incentive, Pandemic Legion offered to keep the peace—for one night only—with any of its rivals who wanted to get in on the destruction of the CFC Titan. Pandemic Legion put out a Call-to-Arms, and the fleets of N3 and TEST Alliance were finally beginning to arrive.

To make matters worse, it was Saturday. Not even working hours could interrupt the ever-growing scale of the battle. CCP was behind the scenes working to keep the servers up.

"Once it was clear that the fight was large, in charge, and not going anywhere, we took the only action that we really have—we moved other solar systems away from that [server] node," said developer CCP Veritas to *EVE* journalist Alizabeth. "Moving a system like this disconnects everyone in it, so moving the fight system itself isn't acceptable, as those in the non-favorable position simply won't log back in. This didn't make a big difference on the performance of the fight, as those systems were mostly empty, but it did at least make those other systems fine after the move."

By sheer chance, a fleet of 60 dreadnoughts from the small-but-mighty mercenary alliance Black Legion also happened to be in the region conducting a fleet exercise when they got the news about what was happening about 10 jumps away. Its fleet commander Elo Knight swiftly made a detour to get those ships in on what was now one of the biggest and most bizarre lowsec battles in memory, and almost everybody's guns were pointed at DaBigRedBoat's Titan.

```
"Throughout the battle, Drunk N Disor-
derly and Lost Obsession's [Heavy In-
terdictors] were key to trapping CFC
ships in Asakai. As Heavy Interdictors
are soft targets for heavily armed cap-
ital and supercapital ships, DnD and
Lost Obsession quickly found themselves
dangerously low on [Heavy Interdictors].
At this point a member of Lost Obsession
devised a clever plan to get more—travel
to Jita, the trading hub of New Eden,
buy as many Heavy Interdictors as they
could stuff into the hold of the fast-
est freighter they could find, and park
them in the neighboring [system named
"Prism."] Before the end of the battle,
DND and Lost Obsession would purchase
every Heavy Interdictor for sale in the
regions around Asakai and in Jita. Ac-
cording to Drunk N Disorderly head Sajuk
Nigarra, they lost well over 20 of them
before the battle was over."
 — Addie Burke, GamingTrend
```

The CFC continued to escalate the battle as well, bringing in many of its own Titans to provide cover for their ships and keep things from turning into an all-out slaughter. But the odds continued to grow in favor of its enemies as ships poured in from around the star cluster.

At the peak of the frantic melee, 2754 players from a litany of more than 270 alliances were in Asakai, one of the largest such events in *EVE Online*'s history and the history of online gaming.

When DaBigRedBoat's Titan became the focus of the entire star cluster, everyone mostly forgot about the worthless starbase that had originally sparked this event. The Liandri Covenant pilots sat safely inside the shield of the tin-can starbase as all hell came to bear on their home system.

Those sheltering pilots must surely have been awed by the power and interconnectedness of this video game community. Liandri Covenant was like a gang of neighborhood toughs whose beef with a slightly larger rival gang escalated into a proxy battle between nation states on their front stoop. No one was even paying attention to Liandri Covenant anymore as the nullsec powers targeted each other's most expensive supercapitals at close range. From inside the bounds of the thin blue shield of their starbase, Liandri Covenant saw a jagged tangle of 3000 ships slipping and lurching through space in a laggy, time-dilated ballet.

As the battle began to slip out of control, the CFC began to pull its ships out one by one, feeding trapped ships enough capacitor power to engage their jump drives and escape the battle.

Pandemic Legion fleet commander Hedliner noted after the battle that had this fight taken place in nullsec instead of lowsec, their Heavy Interdictors would have been able to create warp disruption bubbles that would've destroyed 60% of the CFC fleet. Instead, the CFC managed to evade many of those losses. That's not to say the CFC emerged unscathed, however. Three Titans, including DaBigRedBoat's Leviathan, were destroyed along with 6 supercarriers, 29 carriers, 44 dreadnoughts, and 450+ sub-capital ships.

The near-trillion ISK cost of the battle was far from an existential blow for the ultra-wealthy technetium kingpins, but Asakai was a moment that forced the relationship between the ClusterFuck Coalition to come to a head. After the battle, many of the combatants—including TEST—kept positive standings with Goonswarm and the CFC in spite of the fact they'd just destroyed 850 billion ISK of CFC property. But the Battle of Asakai seems to have put a question in the back of everyone's minds: if we're on the same side, then why weren't we on the same side in the biggest battle of the year?

DaBigRedBoat and the ClusterFuck Coalition as a whole limped away from Asakai. The CFC, an aspirational superpower, had been defeated in a very public battle, witnessed both by the *EVE* community and by the press who had raced to document the unlikely conflict. The battle wasn't geopolitically important, but it became a widely-reported sensation because it encapsulated the specialness of *EVE* into a single story. DaBigRedBoat was temporarily banned from piloting Titans by his CFC contemporaries, who also teased him mercilessly.

Tiberius Stargazer wrote in his retrospective on Asakai: "The battle is the ultimate lesson that even the smallest of rivalries can spiral out of control in the universe of New Eden. Ironically, the [starbase] that was the start of this engagement remained intact until the following morning when it was destroyed by a 3rd party pirate fleet." ●

Above and Below: More screenshots of the Asakai confrontation as it spiralled out of control.

DARKNESS OF THE HONEY BADGERS

"Opinions vary regarding the exact motivations of the split between TEST and Goonswarm. Some believe that TEST's leadership resented the limited way in which the CFC had assisted in the south [against IT Alliance.] Some think, rather, that the seed of the rift lay in TEST CEO Montolio's continued feeling of insecurity and desire to prove himself every bit as much a leader as The Mittani."
— Matterall, commentator and journalist, Talking in Stations

On January 28, 2013 the global news was topped by stories about a deadly fire at a nightclub in Brazil, an oil spill on the Mississippi River, and the Honey Badger Coalition dunking three CFC Titans in Asakai.

The news coming out of Asakai was mind-bending for the average person in early 2013. Video game fans had long been intrigued by what went on inside *EVE*, but mostly because of stories about heists and high-profile deceptions. Those stories were often opaque and difficult for readers to fully grasp. Most of the time you were left with a vague sense that what was happening in *EVE* was awesome even if you didn't fully understand why. People often tried to explain *EVE*, but it rarely worked because *EVE* is steeped in decades of context that nobody outside the universe knew.

But here was a story that anyone could understand. Some dude named DaBigRedBoat misclicked—like any of us have a thousand times—and a cascading sequence of political agreements straight out of a World War I history book caused a ruinous battle. *EVE Online* had a high profile as a major online video game, and it had been highly intriguing to members of the video game community for years, but Asakai put *EVE* on a whole other scale of notoriety.

It's important to remember that even though *EVE* had been creating battles between hundreds and even thousands of players for years before this, the average video gamer wasn't really aware of this except vaguely.

Two-hundred player battles that lasted hours were happening in *EVE* as far back as 2004 when most video games only supported a maximum of 16 players in a single match that usually lasted less than 15 minutes. But the problem with *EVE*, of course, is that everything is happening live, meaning that word of these huge battles usually only reached people days after the fact. CCP had trouble marketing its thousand-player battles because it couldn't promise prospective players they would actually get to participate in one. The best CCP could do was to say that if you play *EVE*, someday the political situation might ensnare you and you might end up in a grand battle. That's exciting in an abstract sort of way, but it made for a tough sales pitch. Asakai, however, was streamed live as it happened, and screenshots were sent to the major gaming blogs in real-time.

But since this was the first time some of these bloggers had ever written about *EVE* before, they tended to over-inflate the importance of the battle in order to make it seem more climactic for readers. This meant that even though Asakai was mostly just a very costly screw-up—a "welp" as *EVE* players would say—many readers were seeing stories about how the CFC was now on its back foot, struggling to recover from the brutality of the battle. The reality was that the CFC's ego was bruised by the bad press more than its supercapital fleet was by the Honey Badgers.

In the *EVE Online* community, onlookers were awed. The battle had been spontaneous, and the objective was mostly to have a good time and kill DaBigRedBoat's Titan. It wasn't about conquering ClusterFuck Coalition space and so the CFC mostly took it in stride and congratulated the other side on a fun event for the players.

Opposite: A fleet manuevers through a battle near an Erebus-class Titan. **Center:** The logo of Darkness.

However, seeing The Mittani or Goonswarm on the losing side of a major battle often has the effect of spurring *EVE Online* players to write near-poetic forum postings. In the wake of the Battle of Asakai one TEST pilot wrote on the internal TEST forums:

```
"After some time, it was clear there
was a general move towards war between
the two coalitions, while the Space
Tyrant [The Mittani] moved to maintain
his precious [technetium] supply above
all else.
    The root of his folly was a failure
to realize his position depended on
the good graces of his powerful al-
lies TEST and Pandemic Legion. [...]
He forgot his true broskis. Whilst
he may claim everything is patched
up, the truth is quite different.
The cat is out of the bag. Mittani's
actions is all about his own gran-
diosity, and he will trample on any
allied coalition, valid issue or not.
CFC priorities are #1; true broskis
are disposable. Make no bones about
this, his own actions have forever
poisoned the well with TEST, and PL
is paying attention too. Who would
want an ally like that?
    The truth is that there is likely
a limited amount of time until small-
er, more skilled pvp alliances start
to capitalize on this situation; alli-
ances with both the anti-goon grudge,
and the expertise to over-match limited
goon skills. It won't take long for the
sharks of EVE to come out and feast upon
the distended whale of the CFC once it
is clear that PL and TEST will not their
savior be.
    Quite simply, the emperor has no
clothes."
    — Anonymous TEST Pilot
    January 27, 2013
```

The relationship between TEST and Goons was fraying as the ambitions of Montolio and The Mittani butted up against one another. Montolio had grown increasingly independent as TEST's power was growing, and it was glorified in the victory at Asakai which had been broadcast in news outlets around the world. Anybody who has played a video game will know how exciting and addictive a thrilling win can be, but imagine if you once won a video game—alongside thousands in your community—on such a scale that it made worldwide news.

That's not just a thrilling video game win. It was perhaps one of the greatest achievements of Montolio's life. It would be for almost anyone. He would later say that after Asakai he coveted another such event, and he knew the only adversary in *EVE* who could deliver it to him was his supposed ally, the ClusterFuck Coalition. No one else could match TEST's numbers to create that level of escalation. So, if Montolio and TEST were going to get another hit of the soaring high of Asakai, they would need to be fighting Goons.

Throughout this time Montolio became increasingly erratic according to multiple sources, as he tried to instigate another battle with the CFC without openly declaring war. Some supported him in this, and others tried to keep the peace. Shadoo of Pandemic Legion, for one, suggested an elaborate war game between the coalitions to blow off steam.

"Several figures both within the [Honey Badger Coalition] and outside the HBC were pressing me to go for full-on warfare with the [ClusterFuck Coalition,]" Montolio later wrote. "This sounded like a great idea to me, casually, because I don't know why everyone else has chosen to play this game, but why I started playing is because of the stories that happen in *EVE Online*. The stories you read about, the wars you read about. The vast coalitions collapsing. Reading about *EVE* is probably more fun than *EVE* most of the time. [...] What I desired was a large-scale conflict. I want to see thousand vs thousand-man battles. Asakai is essentially what I am after."

But he was ultimately unsuccessful at instigating war, because otherwise the state of technetium was too profitable for the major powers. The CFC and Pandemic Legion alike shuddered at the thought of a prolonged war that would cut off technetium production.

Within a short time the stress of the leadership position compelled Montolio to take a break from the game. Throughout February 2013 he remained absent, biding his time and considering TEST's future.

Meanwhile, the ClusterFuck Coalition was busy dealing with its own internal drama. Sion Kumitomo of Goonswarm—who became the director of the Corp Diplomatique after Vile Rat's passing—wrote that on March 6, 2013 he officially notified one of the CFC's strongest PvP groups "Circle-of-Two" that something was so deeply wrong with their finances there was only one conclusion: they were being robbed. According to Sion, Circle-of-Two leader Gigx never took any action based on that intelligence, and would wait another 13 months before firing the director accused of an ongoing theft of hundreds of billions of ISK. It seems the CFC may simply have been too distracted to be drawn in by Montolio's warmongering.

When Montolio returned to TEST Alliance it was only a few weeks before he became overwhelmed by leadership duties and resigned.

"I concede," he wrote. "It isn't what I want, but it is what you want. I want war. I want gigantic fucking battles. I want to crash nodes because people are so fucking interested in this shitty game."

He opened his final post with the common shoulder shrug emoji indicating defeatedness:

```
"¯\_(ツ)_/¯
Shit was too much work, it wasn't fun
anymore. We had friends for strategic
reasons not because we actually liked
them. Internal politics was shit. Run-
ning a space empire was shit. Being on
call for EVE 23/7 was shit.
    I love TEST and I love Dreddit. I
have friends in both. I didn't steal
anything on my way out or disband/kick
anyone. I just couldn't handle the sit-
uation we were in. It is entirely my
fault that we ended up in that situa-
tion, all I can say is that it was fun
up until it wasn't. I would never run a
coalition again, it changes you."
    — Montolio, Executor, TEST Alliance
Please Ignore
        March 21, 2013
```

Montolio's resignation left TEST Alliance without a leader, and after some brief internal disputes the CEO position was taken by a player named BoodaBooda.

SORT DRAGON

That would've made Shadoo of Pandemic Legion the CEO of the Honey Badger Coalition, but Shadoo passed and said he wanted to pull back to spend more time advancing his own alliance's goals. He offered the CEO position to anyone who wanted to give it a shot. In the absence of any obvious takers, a veteran member of Pandemic Legion named Sort Dragon—formerly of Band of Brothers and IT Alliance—offered to take the lead role. However, Sort Dragon had only ever managed individual corporations, never an alliance, let alone a coalition on this scale.

The Australian Sort Dragon formed a new alliance called DARKNESS and began recruiting other corporations to help him form the core of the new Honey Badger Coalition. However, when he began looking into the forums and day-to-day operations of other alliances in the coalition, he was disturbed by what he saw.

"Looking in on this I saw a trend starting to form and I knew something had to be done," Sort Dragon wrote in a post on the TEST forum. "I am the first to admit I myself can be an asshole to an extent, but what I saw was no longer friendly banter or even strong-handed playfulness between friends. What I saw festering in our coalition was a pure hatred for each other and I knew if we were to survive, this could not continue. If we as a coalition were going to stay together we had to find a way to coexist."

Sort Dragon held a meeting with the new CEO of TEST, BoodaBooda, and he laid out what he believed was an existential crisis within TEST. Big changes were needed to the internal structure of TEST or else he believed the alliance would soon collapse. BoodaBooda had often said it was his dream to cut TEST's membership in half and purge the less committed players—essentially what Sort Dragon wanted to do. But at the same time BoodaBooda did not take kindly to the elder Sort Dragon telling him how to operate his brand new alliance. Especially when Sort Dragon's own position of power was somewhat natal. To make matters worse, Boodabooda seems to have gotten the impression that Sort Dragon might try to replace him if he didn't take action.

DARKNESS was far newer and far younger than TEST Alliance. Though Sort Dragon was nominally the leader of the Honey Badger Coalition, it was only a title. Sort

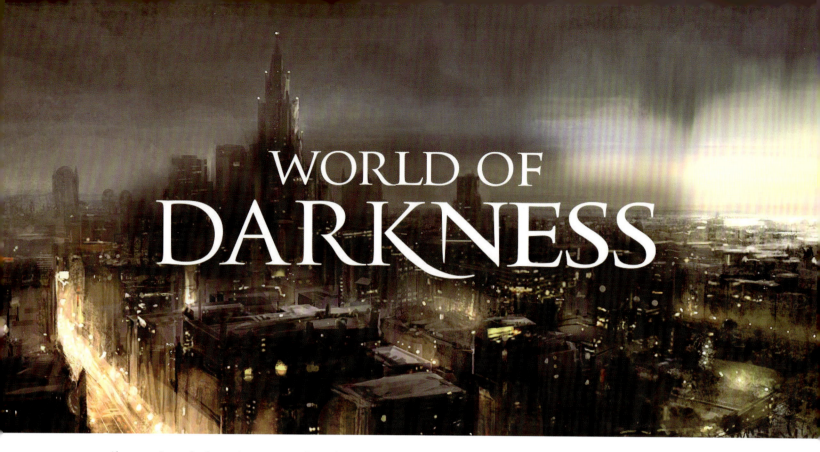

WORLD OF DARKNESS

Dragon and Boodabooda were both just new alliance leaders, and TEST was the far more powerful alliance. Imagine a brand new president of the European Union telling the Chancellor of Germany how to run the country on their first day on the job, and you have a strong allegory. The youthful BoodaBooda—who was only about 20 years old at the time—saw the demands as a sort of threat.

However, all of the drama and realignment in the Honey Badger Coalition was going to have to be put on hold for a week, because in April it was time for all the biggest alliance leaders (and wealthiest alliance members) to pack their bags and head to Reykjavik, Iceland for Fanfest 2013. But not all the drama was safely packed away, and 2013's gathering turned out to be the fatal blow for the powerful coalition.

FANFEST

This was to be the biggest Fanfest yet, a celebration of the tenth anniversary of this amazing alternate reality. With *EVE Online* still climbing to new peaks of popularity, the theme for this year's event was "*EVE*: The Second Decade."

Veterans from the creation of the game itself in the late 1990s came back on stage in the keynote to tell the tale of the creation of *EVE*, and how it all began as an upstart virtual reality company in Reykjavik that didn't know what it was doing. Throughout the event CCP was streaming interviews with the players live to an audience on the streaming platform Twitch.

One night, after the convention had closed, the players were doing what they'd done every night of Fanfest for the past nine Fanfests: cramming into bars. Since the pack of Fanfest attendees is a smorgasbord of nationalities predominantly from around Europe and North America, the group usually sticks to the bars on the tourist-friendly street "Laugevegur" next to the convention center. It's a cobblestone street surrounded by the historic Icelandic parliament buildings that had seen protests in 2011, just down the street from CCP Games' new headquarters building. Since the crash, when Iceland's stint as a banking superpower ended in catastrophe, the Icelandic government had begun investing heavily in modernizing Iceland as a tourist destination. That meant building dozens of new tourist-facing businesses in the downtown area near the convention center.

The historic downtown streets which saw the birth of the world's oldest democratic government were now also lined with gift shops, tour guide centers, and bars sporting American flags and Chuck Norris jokes to attract tourists.

One such small bar played host to more than a dozen *EVE* luminaries that night, drinking along with dozens of others in a noisy bar. Among them were Sort Dragon, his girlfriend, and some of his long-time friends. Late in the evening, ProGodLegend—a rival fleet commander from Nulli Secunda of the N3 Coalition and an old nemesis of Sort Dragon from way back in the IT Alliance days—walked into the bar with some of his friends.

Sort Dragon says that his presence alone was a clear provocation, because the two of them had beef going back years. They were both members of IT Alliance in different capacities, and the resulting split had left them with bad blood that was exacerbated time and again through three years of personal animosity that both of them shared for the other. Though they were nominally allied through the Honey Badger Coalition, neither was fond of the other.

Their enmity for one another grew in parallel to this book's story as time and again they found themselves on opposite sides. It all came to a head that night in the bar. To hear Sort Dragon tell the tale, ProGodLegend became outwardly aggressive toward Sort Dragon's group, goading Sort Dragon to fight and insulting his girlfriend.

CIRCLE OF TWO

ProGodLegend tells the story more like a drunken adventure as might be expected from someone who was in his early twenties and technically on vacation. In an interview ProGodLegend recounted the evening in detail:

"So we go to Fanfest, and we go out to a bar one night. I am hammered drunk, and everyone is hammered drunk. And the group that I was with in Nulli [Secunda] was a bunch of college kids like myself who are a little trolly, a little memey, and just... fucking with people. And Sort Dragon is there with his girlfriend. And he's a little drunk, and he's a little bit of a dick at the time.

So he's at the bar, and the way I remember it is he was being an asshole, and I told him to fuck off. He said, "you wanna step outside?"

I said "sure thing, man, let's do it."

And he didn't want to do it, and I don't think I wanted to do it either,

DUST AND DARKNESS

At its peak, CCP Games was a sprawling global company not only operating *EVE Online* out of its Reykjavik office, but also building the brand new MMO World of Darkness in Atlanta as well as the ambitious and unique EVE-connected multiplayer shooter DUST 514 in its Shanghai office.

But after the Summer of Rage protests, funding for these projects was thrown out of whack amid drastic layoffs across the company. Though DUST 514 was nearly complete by 2013, World of Darkness had experienced drastic delays. Even before the layoffs hit, the Atlanta studio complained frequently that its workers were routinely "borrowed" by the Reykjavik office when *EVE Online* projects weren't making enough progress. Ironically, many *EVE* players viewed these projects as a distraction from development on *EVE Online*, when in fact the opposite was more accurate. Ex-employees from the Atlanta studio said *EVE* was a constant distraction that caused years of delays.

When layoffs came in 2011 the Atlanta studio was the most affected, and World of Darkness became something of a back-burner project. Development continued for years, however, until it was officially canceled in April 2014.

Above: At the 2013 Fanfest CCP also unveiled more information about its upcoming shooter Dust 514 which it boasted would be deeply interwoven with the *EVE Online* universe in fundamanetal ways.

but eventually he and I were starting to tell each other to fuck off, and his girlfriend was right there, and she was kinda standing by him, and she was saying some shit too. And as I'm walking away I said, 'you know what Sort, fuck you and your psycho bitch girlfriend.' I didn't think anything of it.

Next day we all wake up, and we've all got hangovers. People start checking the forums, and someone says to me, 'have you SEEN the ping that Sort Dragon just sent out to the Honey Badger Coalition?' I was like 'no, I have not.'

Sort Dragon [messaged] the ENTIRE Honey Badger Coalition that they were gonna go to war with Nulli Secunda. That shit had gone down at Fanfest, and they were going to go to war with [us.]

And I was like Oh My Fucking God. I can't believe he did that."

— ProGodLegend, Nulli Secunda, N3 Coalition

Sort Dragon's story, which I was asked not to record, comes off a bit more like a horror story in which Pro-GodLegend provoked and insulted Sort Dragon and his girlfriend for no particular reason and then threatened to fight him. The incident left him disgusted and shaken, and he felt he was given little choice.

The next day Sort Dragon was due to give an interview on the FanFest live stream in the wake of his war declaration. ProGodLegend's story continued...

"And so Sort Dragon gets on the Fanfest Live Stream... and it was epic.

You had a CCP person [CCP Guard] running the stream, and Shadoo and Sort Dragon are right there. And [CCP Guard] is asking Sort 'what happened man, why you going to war?"

And [Sort Dragon] is trying not to make it too much about real life, but he's like 'you know, we need a war, war is the best way to gel a coalition and get everyone to work together and you need content and they gave us a reason to do it. And Shadoo is like, 'Yeah Sort. What's the reason?' *laughs*

And Sort Dragon's like 'well some shit went down last night, and they were disrespectful and we need a war.'

And the CCP person is like 'Shadoo, what's Pandemic Legion gonna do?'

And [Shadoo] is like 'we're not really involved in this and we're actually going to be resetting Honey Badger Coalition.'

And that's the first time Sort Dragon had heard of it. The look on Sort Dragon's face when he realized "oh my god, Pandemic Legion is not going to be with me on this' was fucking priceless. Like, Sort Dragon was shaking.

And so everyone goes home from Fanfest, [and] BoodaBooda who is the new executor of TEST was really young. And he didn't like Sort Dragon telling him he needed to make changes or TEST was going to die. And he definitely didn't like Sort Dragon—as he put it—'going to war to defend his girlfriend's honor.' *laughs*

Everyone gets home from Fanfest and BoodaBooda calls an alliance meeting. He's like 'I don't know what the fuck Sort Dragon is doing, he can't tell us what to do. In fact, he can't tell us anything, because we're resetting him. The Honey Badger Coalition is dead. Everyone else in the Honey Badger Coalition is reset immediately. Sort Dragon go fuck yourself.'

This all happened in the course of like four days. It was one of the most amazing moments ever. All the people who were at that bar egging Sort Dragon on were like 'omg we just killed an entire coalition.'

It was great. This was amazing. This was the perfect storm. HBC was a major power player and then it instantly went away."

— ProGodLegend, Nulli Secunda, N3 Alliance, interview excerpt

BoodaBooda came back to his alliance to explain his reaction in a Jabber chat room, saying "we don't want to live under the oppression of a backstabbing tyrant."

The characterization of Sort Dragon as a "backstabber" would become a reputation he had a difficult time shaking—especially given what happens in the next chapter—even though from his perspective his actions were perfectly justified.

BoodaBooda continued in a more public announcement to TEST:

[02:58:07] BoodaBooda > We left the HBC because the leader of the HBC made irrational demands
[02:58:09] BoodaBooda > such as
[02:58:15] BoodaBooda > I will take over your alliance
[02:58:18] BoodaBooda > i will take all your corps
[02:58:20] BoodaBooda > i will take all your space
[02:58:24] BoodaBooda > so i said NO SIR
[02:58:37] BoodaBooda > YOU WILL NOT TAKE MY ALLIANCE
[02:58:42] BoodaBooda > YOU WILL NOT TAKE MY CORPORATIONS
[02:58:45] BoodaBooda > YOU WILL NOT TAKE MY SPACE
[02:58:52] BoodaBooda > YOU WILL NOT MAKE DEMANDS OF TEST ALLIANCE
[02:58:56] BoodaBooda > WE WILL NOT GO QUIETLY INTO THE NIGHIT
[02:59:01] BoodaBooda > WE WILL NOT GO DOWN WITHOUT A FIGHT
[02:59:04] BoodaBooda > WE'RE GOING TO LIVE ON
[02:59:07] BoodaBooda > WE'RE GOING TO SURVIVE
[02:59:10] BoodaBooda > TODAY WILL BE
[02:59:16] BoodaBooda > OUT INDEPENDENCE DAY
[02:59:19] BoodaBooda > R*

Sort Dragon, for his part, claimed the war was a necessity, the incident at the bar had merely made ProGodLegend and Nulli Secunda the target.

"The whole thing about my partner being called 'A Psychotic Dumb Bitch' by ProGodLegend more than once without cause just gives me personally something to get behind," he wrote.

As BoodaBooda declared TEST's independence he also gathered the alliance for a "State of the Alliance" speech which he summarized in a blog post:

"A week or so ago, I took over as the
TEST executor. So far I'm having a great
time, really. This leading thing isn't
so bad. […]

The huge critical point from the
[State of the Alliance speech] was the
declaration that we were rebelling from
the HBC empire. This is the juicy bit
that I had to hide from everyone for a
week. Tearing down the house Montolio
built was actually incredibly fun.

I took TEST executor and began getting involved during Fanfest. I was surprised when we started getting information leaked from Fanfest about Sort's
plans for the HBC. […] When Sort finally
came home, he had us all waiting with
baited breath. No one really knew what
he was going to say. At this point most
of us were very upset with the whole
situation. […]

Sort confirmed all the crazy rumors
we heard. […] He explicitly told me how
I needed to change TEST to better com-
ply with his demands. He threatened us,
saying TEST was doomed to fail. […] He
told me that if Sort Dragon says we do
something, we do it without question or
hesitation. […]

I could go on and on with this, but
the general idea is that Sort showed an
absolutely critical lack of leadership
ability or understanding of how TEST and
the HBC function.

For the next few days I had to work
hard to keep things pretty quiet. I
discretely convened most TEST CEOs together to reassure them that we had a
plan, and that I was going to make *EVE*
more exciting for us than it has been in
years. I strategically leaked information to insecure channels so Sort would
catch wind. […]

What I'm saying is, we are an alliance

with ABSOLUTELY NOTHING TO LOSE. Everything up to and including getting brutally
murdered would be a great opportunity to
do something new.

When a threat comes knocking on our
door, we will face it with our historic reckless brutality, and the force of
thousands of screaming, suicidal pilots.

We will defend our home with every
last POS on our moons, every last SHIP
in our hangars, and every last god damned
ISK in our wallets. […]

So, who's with me?"

— BoodaBooda, Executor, TEST Alliance
Please Ignore

THE CFC AND PANDEMIC LEGION

As the drama tore apart the Honey Badger Coalition, Pandemic Legion was already long gone, spending
most of its time out in the Drone Regions attacking
Solar Fleet. It was believed that the reclusive Russians
had grown weak after months of idle land-holding as
the only viable Russian alliance out there. Many of
its players—Mactep included—became more and more
removed from the game. Pandemic Legion was looking
to establish its own renter farm to fund its high-stakes,
high-cost supercapital gameplay, and the Drone Regions
proved easy pickings.

To the surprise of few in the *EVE* community at the
time, PL and N3 (Nulli Secunda, NCdot, and Nexus
Fleet) carved through the massive and largely vacant Solar
Fleet renter empire quickly, riven as it was in the wake
of the BigMaman defection, the Mactep revelations and
the Russian Civil War. Within only a couple of months,
Pandemic Legion had created for itself a huge plot of
rentable territory while sending Solar scrambling back
to empire space to regroup.

When its fleets arrived to take ownership of the
hundreds of Drone Region stars they often found
average players flying through star systems that once
belonged to Solar Fleet. When Pandemic Legion
destroyed those vessels, fleet commander Elise Randolph once told me they'd often get a chat request

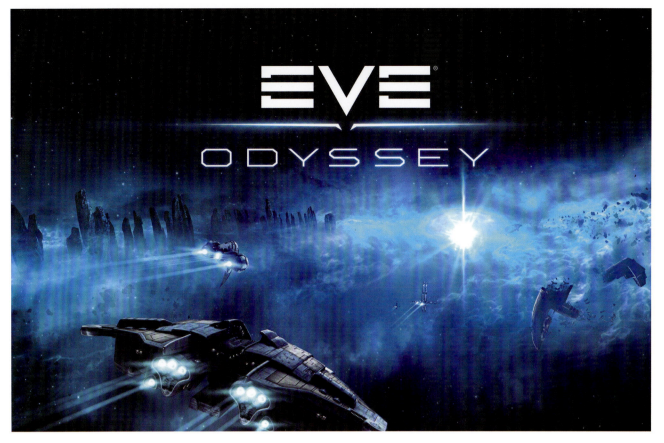

Above: The Odyssey patch was *EVE Online*'s 19th free content expansion. **Below from left to right:** Sort Dragon of Darkness, ProGodLegend of Nulli Secunda, and BoodaBooda of TEST Alliance.

from the person they'd just destroyed. But these players weren't upset or looking to vent at the group that destroyed them. Instead they were friendly and humble. They'd say things like, "whoops, my fault, I didn't realize the space had been conquered. Who can I talk to about renting from you guys? What's the Paypal link?"

According to multiple unconfirmed stories I've been told, the former renters of the Drone Regions were so accustomed to paying for access to their space in real cash that they didn't actually know it was against the rules of the game. In fact, they were often annoyed when

Pandemic Legion told them they had to pay in ISK, not cash. They'd grumble about how it was way more annoying to do it that way.

As Solar Fleet collapsed, Pandemic Legion and N3 emerged as perhaps the only counterbalance still left to combat the ClusterFuck Coalition. Many of the enemies of the CFC were still the same individuals that they had always been, veterans of the Great War and of the alliances the CFC had managed to destroy over the years. As I've written in previous chapters, individual players can't be destroyed by the wars they participate in—the only true casualty is the phenomenon of their cooperation. There

was a sense that, if only the enemies of the CFC could work together, they could shake up the power structure of nullsec. But with the drama so thick after Fanfest 2013, opportunities for cooperation were in diminishing supply.

BOTLRD

The dissolution of the Honey Badger Coalition left Pandemic Legion with no serious way to existentially threaten the CFC, and yet the CFC still had a tremendous amount to lose if Pandemic Legion was to commit to an organized harassment campaign against the CFC's renters and moons. With the geopolitical situation at a stalemate, the two saw a benefit in keeping things civil so they could continue earning enormous sums of money.

The standing agreement between Pandemic Legion and the ClusterFuck Coalition was that neither superpower would challenge the sovereignty of star systems that were owned by the opposing bloc. It wasn't necessarily an alliance. Nor could it be called a non-aggression pact because they were not agreeing to abstain from aggression entirely. They only agreed that the renters and their systems wouldn't be disrupted. If either broke the territorial truce it would result in mutually assured destruction, because reliable two-way peace is required to make broad renter programs work. As long as the money kept rolling in, the two had similar enough values to tolerate one another.

The agreement became known as the B0TLRD Accords, a portmanteau of the initials of Pandemic Legion's renter corporation "Brothers of Tangra" and "PBLRD" (pron: pub lord) which was the commonly used abbreviation of the CFC's coalition-wide rental program also known as the "Greater Co-Prosperity Sphere."

However, even the ostensibly agreeable cause of making money for the entire coalition became a source of friction inside the CFC. According to the new top diplomat, Sion Kumitomo, the player "Gigx" of the CFC alliance "Circle-of-Two" was now demanding that renter territory be given to his alliance so he could attract new recruits. Gigx was envious of the kinds of social programs other wealthy alliances in the CFC had. Sion clapped back that Gigx's poverty was not because the coalition hadn't given him enough, but because—as Sion had already warned

him—Gigx's own financial director had been robbing the alliance blind of hundreds of billions in ISK for months.

"We repeatedly denied CO2's requests for more space in [the freshly conquered] Vale of the Silent since it was being used for coalition income," wrote Sion Kumitomo in 2016. "[Circle-of-Two] advocated that their personal income as an alliance mattered more than shared income across the coalition, and we politely replied that the entire coalition took that space, and should benefit in the best way possible. CO2 spent months furious about this. Note how greed, isk, and more of both are becoming themes here?"

Though it was invisible for the rest of *EVE* to see, the CFC was dealing with serious internal issues that were threatening the coalition's future.

New Eden's nullsec territories were essentially evenly divided between the CFC and the coalition between the N3 alliances and Pandemic Legion, and two of those entities had made a pact not to attack one another. Most of that massive territory was either vacant or renter space. For the time being, a chaotic peace reigned in nullsec as the wealthy blocs could not attack each other, and assuaged their boredom by picking on smaller entities.

The only hiccup in this simplification of the star cluster was that TEST was growing increasingly independent, and through numerous political actions had sought to distance itself from the ClusterFuck Coalition which had been its ally from TEST's earliest days. BoodaBooda may have been taking TEST in his own direction, but he inherited Montolio's decisions and his enemies, and had already made some of his own. With the ink now dry on the B0TLRD Accords, opportunities for shooting stuff were becoming scarce on both sides, and TEST's "independence" was now looking like a liability. BoodaBooda once wrote during this time that he was getting the sense that The Mittani was trying to trick him or some TEST member into breaking some sort of rule. He said he felt like The Mittani was probing for some cause to pin a public uproar upon.

But the trigger that would set off the war was not a scandal but a patch. CCP was beta testing changes to the game which were poised to break up the CFC's technetium cartle—the Organization of Technetium Exporting

Corporations (OTEC)—which formed the backbone of the finances of the CFC. Those changes—part of the Summer 2013 expansion "Odyssey"—were poised to move that money-making potential into Fountain and Delve, both owned by TEST.

ODYSSEY

"In June, CCP tweaked the economy with the introduction of Alchemy," wrote the journalist, history writer, and podcast host Matterall in a retrospective. "Alchemy allowed other moon materials to replace technetium, breaking tech holders' monopolies. In addition, new moons were seeded, including a high number of valuable moons in Fountain and Delve."

The Mittani quickly realized that, if he did not act immediately, Goonswarm's home in Deklein would become massively less profitable, while TEST would get much richer. Worse, some stronger entity might remove TEST and take the moons for themselves, becoming a wealthy border threat. The Odyssey expansion was set to launch on June 4th, 2013, and when that day came the CFC would have choices to make. CFC leadership at first tried to gain control of the moons through diplomacy, but the negotiations fell through leaving the CFC to worry about its future.

The Mittani weighed his options, and gathered the alliances of the ClusterFuck Coalition for an historic State of the Goonion address where he would reveal his plan for the future of the CFC and nullsec broadly.

Meanwhile, Sort Dragon and DARKNESS were left without allies, but that didn't mean Sort Dragon had no role left to play. He never forgave BoodaBooda for dooming the Honey Badger Coalition just as he took control. As The Mittani prepared to deliver his speech, Sort Dragon seethed. His retaliation would come swiftly, and TEST would say it was yet another Sort Dragon "backstabbing."

"It wasn't backstabbing," Sort Dragon told me. "It was retribution." ●

Opposite: A new alliance called Brave Newbies formed and grew large around this time period though its story would not intersect in a major way with nullsec events for another year or more. Below: The dark side of a ringed planetary system silhouetted by its star.

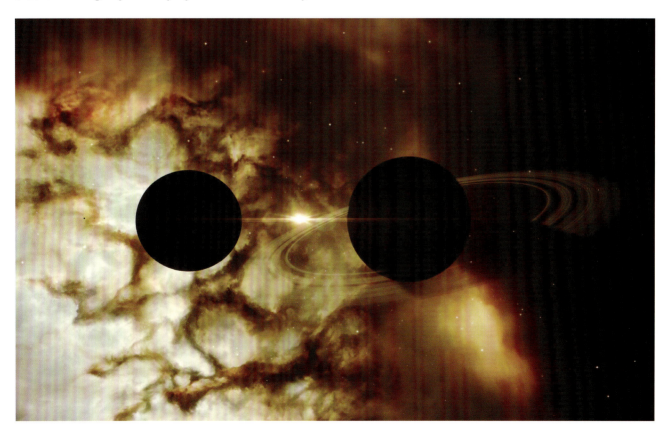

THE FOUNTAIN WAR

"[The Odyssey expansion] has come, and it has left us destitute...Who is right and who is wrong doesn't really matter: if we do not invade Fountain, someone else will—and they will enjoy the kind of wealth that we once had, while we wallow in wrenching poverty."
— The Mittani, CEO Goonswarm Federation

On June 6, 2013, a pilot named Trii Seo of the nullsec corporation "Executive Outcomes"—a member of the ClusterFuck Coalition—was sitting in a cloaked Tengu scanning for data sites. It's an exploratory process that sees players deploying probes to scan for "cosmic signatures." After pinging the data site, Trii approached a small structure in deep space—a solitary husk of technology of indeterminate purpose with gyroscope-like metal rings rotating around it—underneath the famous Cloud Ring nebula.

"One of the most breathtaking of [New Eden's] nebulae is the Cloud Ring," wrote *EVE* journalist Lee Yancy. "Upon approaching the Cloud Ring region in the Western Rim of the galaxy, the roughly circular nebula that the region gets its name from appears and begins to grow larger as your ship travels across dozens of light-years of space. Once a ship enters the region, the pale green sea of gas suddenly dominates the sky. The ring-shaped nebula can be seen cradling star systems in almost every direction, providing one of the best examples of *EVE's* dynamic space scenery."

As Trii Seo finished up the little mini-game you have to play to "bypass the defense subsystems" of the data site and collect your reward, he noticed on his overview that the local population within the system had suddenly surged from just 2 to over 200 ships. Something was up. Not that it was anything to be concerned about, because all 200 of the ships were allies. It was a friendly fleet moving through Cloud Ring on its way somewhere. Trii had no

idea what they were up to, but it must've been important to warrant getting this many people in fleet.

In awe at the sight of hundreds of alliance-mates, Trii warped from the relic site in deep space to the stargate the fleet was arriving through near this system's second planet. Not many people know what it's like to see a warfleet like this in the wild, let alone by accident. Many people only see *EVE* fleets when they're bogged down by time dilation and lag in the major fleet engagements that make the news. Feeling the weight of the fleet in full operation is another thing entirely.

Trii swelled with pride upon viewing the huge formation of 200+ friendlies. Dozens upon dozens of ships in this convoy continued to rip through the stargate in beams of light and massed on the other side. After taking a quick peek around, weaving through the spherical formation and using the "Look At" command to send the camera to inspect the coolest ships, Trii Seo logged in to Jabber to see what was going on.

"Warping to the local jump bridge I see allied ships landing by the dozens," Trii wrote. "Then a Jabber broadcast from our directorate tells us to tune into the coalition Mumble server, because we, as Executive Outcomes, are going to back the CFC in a war."

The target was TEST Alliance's Fountain region. Trii was told to pack up a Black Ops cloaking ship, and get ready to go behind enemy lines. Moments earlier, The Mittani had concluded a State of the Goonion address to more than 2000 subordinates on Mumble, announcing the beginning of the operation.

"Welcome to our fellow Goons, our CFC coalition partners, our frenemies, our temporary allies of convenience, and even to our enemies—thank you for coming today. I'll keep it brief.

My people, Odyssey has come—and it has left us destitute.

For the past few years our coalition has become accustomed to a lifestyle that is the envy of *EVE*. Due to technetium and sound fiscal management, we have been able to fund a peacetime reimbursement system, carrier giveaways, and galaxy-changing events like Burn Jita and the Ice Interdictions. But now that has all changed. [...]

Wars in *EVE*—and in the real world—begin with a declaration of grievance, followed by a quest for justice. The enemy did a bad thing to us, and now we must go after them to redress the balance; they are bad people and need to be punished. White Noise DDOS'd our servers, so we took Branch. Raiden abused Titan blobs against our people, so we took Tenal. NCdot violated a treaty, so we took Tribute. So it goes.

Just as the aggressor claims a grievance against the defender and demonizes them, the defender inevitably claims that the reasons stated for the war are a lie, and that the motives for the war are a crass manipulation of politics. We are being attacked just for money, for space, for greed, say the defenders: the aggressors are lying shills. We have seen this drama played out for every war, on every [forum thread.]

Tonight we are invading Fountain. But we are not going to bother with stating a grievance or demonizing the defender—we are doing it because we need the region and its moons for our people and our friends.

TEST has abruptly found itself in an untenable position as a sovereignty holding alliance—in its current space, at least. Odyssey has vastly enriched the Southwest, while simultaneously there have been a series of unfortunate errors which has alienated most of their former allies, many of whom are now actively attacking them. [...]

What is happening now in the Southwest is a mass land grab motivated by economic pressure and diplomatic realignment: the owners of the richest space in the game have managed to politically isolate themselves and turn their strongest allies into enemies in a matter of weeks. Meanwhile, the Russian forces in the southeast are putting pressure on the former HBC to find new homes in the southwest.

Who is right and who is wrong doesn't really matter: if we do not invade Fountain, someone else will—and they will enjoy the kind of wealth that we once had, while we wallow in wrenching poverty. [...]

Now we go to war. There will be no Q&A session. Every node en route to the staging area has been reinforced to make way for our coming. Our goal is now to get the CFC staged and commence siegework as soon as possible. Ops will continue into the late US time zone.

One final note: do not underestimate the defenders. This may be a long war, and we have grown soft from too much peace."

— **The Mittani**, Goonswarm Federation June 6, 2013

The Mittani was essentially making the case that if the CFC didn't displace TEST Alliance, then TEST's own troubled diplomacy would mean that someone else would. He was framing this not as a betrayal or even as

202

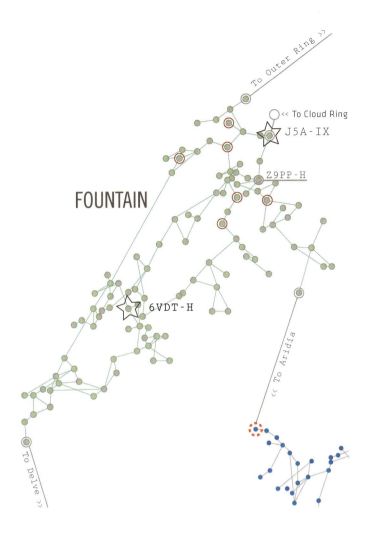

To Outer Ring >>

<< To Cloud Ring

J5A-IX

Z9PP-H

FOUNTAIN

6VDT-H

<< To Aridia

To Delve >>

GOING TO KILL THE FUCK OUT OF THESE NERDS IN FOUNTAIN, AND THEN WE'RE GOING TO KILL THE FUCK OUT OF THESE NERDS IN DEKLEIN."

In some ways it was a typical response from Booda-Booda. He was a leader raised up by Reddit itself, and he wielded power in a typically Reddit-style. Here we see him not so much leading as much as he was trying to start an avalanche. He rallied the anonymous masses and summoned waves of emotion to put hundreds of pilots in fleet. He faced a similar problem other leaders of New Eden had confronted over the years. He needed to convince TEST's pilots to show up and commit to a fight that would be long, arduous, and—given that it was against the most powerful bloc in nullsec—possibly hopeless. He did so by drawing on raw emotion and giving them the feeling of taking part in something larger than themselves. He also did it by promising that no matter what happened TEST would go down in a blaze of glory that they'd be proud to be a part of.

If The Mittani wanted to take Fountain he would need to fight through a classic threadnought given form in starships.

THE INVASION OF FOUNTAIN

Even as The Mittani announced the invasion of TEST Alliance Please Ignore, CFC fleets were already being mobilized. TEST leadership had been aware that the war announcement was in the works for weeks already and had spent that time squabbling about how to respond. Now the ClusterFuck was coming whether TEST was prepared or not.

The CFC main fleet was heading for the border of Fountain in a system called J5A-IX, where it hoped to establish a beachhead into the region. Meanwhile, Trii Seo and his corporation Executive Outcomes were ordered to deploy their cloaking guerilla ships behind enemy lines in the core of the Fountain region and cause all kinds of logistics hell.

Trii and the small gang would sit cloaked in TEST space and count the number of ships in the TEST fleets as they

an act of hostility, but more of an unfortunate real politik necessity. He was also seemingly trying to solidify nullsec's vision of TEST as a lonely, isolated power. TEST still had friends, even if things were strained.

As TEST Alliance leadership scrambled to arrange a response, it was hampered by a mysterious DDOS attack on its voice chat platform.

"Shortly before Goonswarm's State of the Alliance earlier today, TEST [voice comms] fell victim to a SYN flood-style DDoS attack, and has been in disarray since," reads an article by author 'Parliament' on TheMittani.com. "Lack of [voice chat] made the initial response to the CFC's invasion announcement one of disorganization."

While the State of the Alliance was being prepared, TEST leaders posted in Reddit threads instead.

"TEST isn't just a conglomerate of random pubbies," wrote TEST's Military Director Beffah in a Reddit thread. "If the now horribly destitute goons want to take what's ours, we'll make them bleed every step of the way."

Posting in the same thread, TEST leader BoodaBooda offered a somewhat less measured response, "WE ARE headed off for the first battles along the border. Though the cloaked scouts were separated from the main force, they'd try to keep track of how well the battles were going by counting the number of TEST escape pods that floated back. The Fountain War reached them mostly through chatting with friends on the front lines and listening to leadership speeches. Their day-to-day job was to make life as difficult and time consuming as possible, by sniping TEST's newbies, disrupting money-making operations, and generally forcing TEST to do everything the hard, annoying way so they'd log out and go play something more fun.

They had plenty of work to do, and it was growing more and more important because TEST had managed to convince its larger supercapital friends to help defend Fountain. Nulli Secunda and the rest of N3 viewed the strike against TEST as an attack on all anti-CFC alliances during a moment of weakness, and declared that it would help TEST hold Fountain. Pandemic Legion recognized the seriousness of the situation, but responded in a typically aloof way. PL insisted that it would only be in Fountain as an "unaffiliated third party." PL was there to shoot the CFC because it was a clever loophole in the B0TLRD Accords, (Pandemic Legion never said it wouldn't attack the CFC, just that it wouldn't disrupt its renter operation.)

With the supercapital fleets of N3 and Pandemic Legion now committed, the ClusterFuck Coalition's main fleet was having a difficult time cracking through the critical system of J5A-IX.

A DRAGON SLEEPS IN FOUNTAIN

However, TEST had a weak point in the border systems that only those in top leadership positions fully understood and feared: several of the systems on the border were technically owned by Sort Dragon through a subsidiary, a paperwork technicality leftover from their old coalition relationship.

"On June 9, 2013, TEST abruptly lost control of five systems deep within Fountain," wrote gaming journalist Addie Burke. "Under a special arrangement, Sort Dragon had been allowed to maintain control of these systems when TEST left the [Honey Badger Coalition.] Apparently bearing TEST ill will over the HBC's demise, Sort

Dragon abandoned control of 4-EP12, PXF-RF, R-BGSU, XUW-3X, and ZUE-NS—but not before alerting the The Mittani of his plan."

The ClusterFuck Coalition black ops fleets had already been moved into position to harass some of these five systems within Fountain which were very suddenly up for grabs. Within hours the CFC had captured four of the five systems. It wasn't exactly a foothold, since the systems were deep in Fountain and isolated from the frontline, but they represented an annoyance to the defenders who now had to divert time and resources to get these systems back under control.

The very same day, June 9, 2013, with TEST reeling and trying to recover the lost systems, the CFC main fleet struck the gateway system into Fountain with full force.

THE BATTLE OF J5A-IX

"Three CFC fleets attacked TEST infrastructure in J5A-IX in an attempt to capture the system," wrote Burke. "As the single connecting point between Fountain and CFC home territory, capturing J5A-IX was an essential step in any planned invasion. If Allied (TEST/N3/PL) forces could hold the J5A-IX choke-point, they could deny the CFC a beachhead in Fountain."

However, the loss of five systems uncontested ignited a furor in TEST Alliance over what they perceived to be another Sort Dragon betrayal. Sort Dragon saw it as comeuppance, but the memes were merciless.

TEST members were suddenly even more inspired to get online and hold the line in J5A-IX. With TEST's numbers suddenly invigorated, BoodaBooda finally got the avalanche he was looking for and it pushed the CFC fleet out of the system.

In a moment of celebratory grandeur, the TEST/N3/PL combined fleet chased the CFC out of J5A-IX with overwhelming force. Drunk on victory, they pursued the CFC across the border and back to the station the attackers were staging from. The pursuit turned out to be wildly overzealous however, and it was broken by the CFC's near-endless ability to re-ship its pilots outside its own supply depot.

But avalanches of enthusiasm don't last forever. The two sides continued to skirmish in J5A-IX for some time

Above and Below: Fleets of attack ships flock in unison. Oppposite: The logo of Black Legion, one of the most capable third-party mercenary fleets still operating in New Eden.

afterwards, but after five days the strain of trying to hold the system while recapturing the Sort Dragon systems finally became too much. TEST lost control of J5A-IX, the bridge into Fountain, and the CFC surged forward into the breach.

At this point, the war had become huge news within the *EVE* community, and hundreds of thousands in the wider gaming community were now aware of the public falling out between TEST and Goonswarm, one of 2013's great internet dramas. This was great for recruiting pilots to fight in the war, but it also attracted opportunistic pirates who thought this might be the perfect moment to take their revenge on distracted nullsec enemies. While TEST was fighting at J5A-IX on the northern tip of its empire, TEST's resource-generating moons in the far south came under attack by a group calling itself Confederation of xXPIZZAXx.

Similarly, the CFC's richest moons in the far north came under attack by the veteran dreadnought fleets of Black Legion and its lead Fleet Commander Elo Knight (a rare third party force in *EVE* capable of using capital ships to strike the biggest blocs and living to tell the tale.) Rather than diverting forces to defend against Elo Knight, however, the CFC instead offered Elo Knight a contract as a mercenary to move south and harass TEST instead.

"The war was not going well in the south (Fountain,) and bouncing back and forth was severely hurting our ability to either defend our space or prosecute the war," wrote Goonswarm's Sion Kumitomo about the Elo Knight attacks. "After Elo Knight wiped out our [defense] fleet, we signed a [mercenary] contract with Black Legion, effectively eliminating the threat to our back yard, and allowing us to focus on Fountain completely. This was important as we were still very much outnumbered in Fountain, and losing ground."

Though they had already damaged CFC space, Elo Knight and Black Legion became an important arrow in the CFC's quiver. But the space that was damaged in the initial attack became a source of administrative friction in the CFC.

"After the contract was signed, Gigx pulled [Circle-of-Two] back to 'eliminate hostile towers' in the north, which were of course the old staging towers [Elo Knight] had been using to hit our holdings. With [Elo Knight] under contract, there was no threat from those staging towers, but Gigx temporarily pulled his forces from the war—and jeopardized the coalition's chances in the war and our newly signed [mercenary] contract with [Elo

Knight]—to appease his pride. This was the first sign that [Gigx's] mentality had started to shift away from 'how can we help the coalition and prove ourselves' to 'fuck you, got mine.'"

The strains of wartime squeezed and pressured both coalitions in unforeseen ways, and developed cracks that wouldn't be fully understood until years later. One of the most difficult tasks of coalition leaders in *EVE* is managing dozens of egos and conflicting interests, and in this case a resurgent Gigx was feeling less and less beholden to the larger ClusterFuck Coalition.

HAARGOTH AND LOVEGOOD

Sometime in late June, The Mittani was approached by a pilot named Xenophilius Lovegood. Xenophilius was a director in the rental association run by Nulli Secunda which was called "S2N Citizens." S2N Citizens was the shell alliance through which Nulli Secunda rented out its space. That rental money constituted a critical income source for funding N3's presence in the Fountain War. Lovegood approached The Mittani with an offer that was now practically routine within *EVE Online*. Steal. Disband. Defect.

The Mittani saw an opportunity to deal an existential blow to his ancient enemies in N3. Just as spywork had undone his foes in the Great War, so would it dispatch the imposing Titan fleet of ProGodLegend's Nulli Secunda and NCdot (who still counted such luminaries as SirMolle among its membership.) But this was a delicate operation that required a very specific sequence of User Interface actions in order to properly exploit. So The Mittani called in the only person who had ever personally executed a heist and disbandment on this sort of scale: Haargoth Agamar, the Traitor of BoB. Although branded with the mark of the traitor, he'd been serving dutifully within his new alliance as a diplomat under the very same Goonswarm Intelligence Agency handlers who had helped him defect to Goonswarm back in 2009.

Agamar showed Lovegood exactly what to do, and Lovegood carried out the heist on July 1, 2013. The next morning, the holding corporation "S2N Citizens" ceased to exist, and N3's sovereignty dropped in more than 200 of its star systems. Its entire rental empire, which brought in hundreds of billions of ISK per month, was shut down, and a free-for-all began as smaller alliances rushed to grab the former holdings.

N3 announced its immediate intent to pull back from the Fountain front to reposition its fleet and recapture the 200 lost systems in the south. Pandemic Legion would join them to help shore up that critical income source. The effects on the battlefield in Fountain were felt immediately. N3/PL's fleet of Titans and supercarriers were most effective in their role as a bully that could scare off the CFC's subcapital fleets. N3/PL's Titans often won battles simply by being in position and preventing the battle from ever taking place. The CFC had now managed to completely isolate TEST from its allies. Without N3 and PL's supercapitals—and the critical battlefield intimidation factor they provided—the situation for TEST began to deteriorate.

As small battles began to tip toward the CFC, the CFC "Sky Command" team grew more and more comfortable deploying its own heavy ships toward the front line. Momentum was growing, and players were excited that the slog was finally paying off. Plus, they were now flying in fun operations that weren't quashed immediately by the N3/PL Titan fleet. Eventually, TEST was forced to field its own capital ships to fill that void.

BALTEC FLEET

Throughout the war thus far, the CFC had been experimenting with a number of different fleet archetypes to crack the defense in Fountain. Everyone was required to fly in a certified ship or else it wouldn't be reimbursed if destroyed. One pilot, however, refused the orders every time, and insisted on only flying his favorite ship class: the Megathron Battleship, which hadn't been in regular use for years. No matter what kind of fleet composition was ordered, the pilot "Baltec1" showed up in his signature Megathron, earning unanimous ire from CFC fleet commanders. But because he always showed up on-time and paid attention, this strange obsession was indulged. Baltec1 tinkered with his favorite ship's component configuration relentlessly, determined to unlock the inner secrets of the Megathron's capabilities.

BATTLESHIP // GALLENTE FEDERATION
HEAVY DAMAGE
MEGATHRON

LARGE HYBRID WEAPONS

"[Baltec] made fame by showing up to every single fleet in a Megathron," Goon fleet commander Mister Vee told PC Gamer reporter Steven Messner. "People were not encouraged to do silly stuff, everyone had to shut up and get in line. Baltec, however, always stood out. He was so stubborn."

"Until Baltec showed up, it was stupid to think that a Megathron would ever do something different," wrote Messner, one of the most prolific chroniclers of New Eden in the gaming press. "But Baltec is a theorycrafter. What others saw as impossible, he saw as an exciting challenge. He began spending all his time and ISK fitting his Megathron with modules that would help bridge the gap between it and the ships he was flying with."

But over time, Mister Vee started to notice that even when his fleets were decimated on the Fountain border, Baltec1's Megathron always seemed to survive.

"It started building this reputation as an unkillable ship because it kept coming back," Baltec1 told PC Gamer's Messner. "It was like a morale boost. When you lose fleet after fleet after fleet, people stop showing up. But when this thing was coming back every time, it gave people something to celebrate."

When Mister Vee and his theorycraft advisers were informed that the war was costing too much money for the cash-strapped CFC, they were tasked with coming up with a fleet concept that made sense in the high-casualty warzone they now found themselves in. The previous fleets were good, but they were also expensive, which isn't great in fights with hundreds of casualties. As the war began to drag, the Sky Command team realized they needed something cheaper, more survivable, and above all, something the membership would enjoy flying.

That's when Mister Vee remembered Baltec1. On paper, Baltec1's unique Megathron—honed to perfection over months of experimentation—was exactly what the CFC needed. With his configuration it was cheap, tough, fast, and the CFC could afford thousands of them. Vee gave the order to his pilots to begin training for "Baltecfleet."

Above, Below, and Opposite: The Megathron-class hull came back into style in Goonswarm fleets during this era thanks largely to the theorycraft efforts of lone believer Baltec1.

THE MIRACLE OF Q9PP-H

The CFC still controlled only two systems inside Fountain—one of the former Sort Dragon systems alongside the freshly conquered J5A-IX—and TEST was constantly trying to push its adversary back out of the region at every opportunity. TEST wanted to clamp down on the chokepoint stargate pipeline that led into Fountain, so one day it anchored defensive hardware in the midpoint system along the route.

"It was a day like any other for the combatants in Fountain," wrote CFC fleet commander Kcolor.

The CFC needed that pipeline to stay open if it was going to have any hope of breaking Fountain, and CFC fleet commander Mister Vee responded by attacking the defensive emplacements with his brand new Baltecfleet before the emplacements were fully online. TEST didn't want to be pushed around by a bunch of Megathrons, however, and it began preparing to retake battlefield dominance.

As it would happen, this was the day before TEST Alliance had scheduled an in-person gathering of the North American TEST membership in Chicago, Illinois. BoodaBooda was on his way there while this was happening, and Military Director Beffah took the role of delivering a State of the Alliance to focus attention on what was happening in Z9PP-H. Hundreds of TEST pilots gathered on Mumble, or listened the next morning on SoundCloud.

On the recording, Beffah can be heard delivering her speech while also swatting down bad suggestions from the alliance chat room for the upcoming propaganda Theme Week.

```
"Tonight—after we're done with this—
we're going to go save an iHub in
Z9PP-H, and we're going to kick some
Goons asses. There's like 500 of them
in local right now in [their staging
system] B-D. This entire weekend we
have some major major major [stations
coming out of reinforced mode.] We're
coming to the final timers of all the
sov that dropped when Sort dropped it.
I know [the timers] are really really
early for most of us - it's like 4 in
the morning my time.

    There are going to be a lot of peo-
ple away for the Testival so Euros this
```

```
is your time to shine. I know a lot of
you aren't listening to this live and
you're listening to it in the morning on
a relay, but this is your time to shine.
We have the numbers to send Goons back
to Cloud Ring, and this is where you
need to show up. Participation. Fleet
cohesion. I know you guys like flying
whatever the hell it is you fly, but if
the [Fleet Commander] calls for Foxcats,
show up with a Foxcat ship. Don't be that
dude who shows up in a Rifter. If that's
all you can fly, show up in a Rifter,
but if at all you can: show up in a DPS
battleship, show up in a logistics ship,
show up in a webbing Loki. Booster pi-
lots, I know it sucks because you don't
get Killmails, but we absolutely cannot
run fleets without you.

    Grr Goons.

    I have one request for you nerds. We
need propaganda. Good propaganda. Show
up in fleets and make propaganda. […] I
know there are dozens of you nerds who
are amazing with Photoshop and Paint.
Post all of your shit.

    Like I said I'm making it short and
to the point because the Z9PP-H timer
is in about 40 minutes.

    — Beffah, Military Director, TEST
```

The meeting was interrupted when a TEST member suggested an Ayn Rand-theme for the Propaganda Week, and the meeting devolved into a fight about Libertarianism. It eventually recovered and a fleet was formed to head out to Z9PP-H. It wasn't long after arriving that the fleet met Mister Vee's Baltec fleet.

The CFC leadership logged on and spammed the chat channel to rally the membership to rendezvous with a fleet and head to the battle site. TEST responded by calling up carriers and dreadnoughts to try to gain an overwhelming advantage. Manfred Sideous of Pandemic Legion offered to help out with a fleet of his arguably-impossible-to-kill "Slowcats." However, without sufficient support ships on the battlefield, this wing of the carrier fleet was scrambled as well, and a post-battle report by Manfred Sideous suggests his fleet and TEST's were in such disarray that

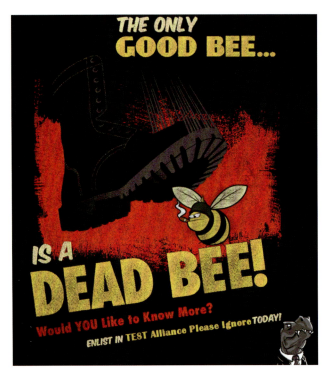

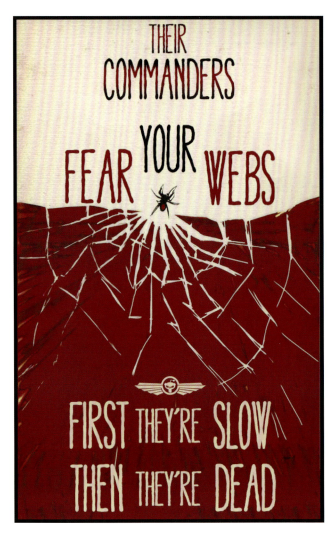

Above and Right: Propaganda posters created by TEST Alliance members. The poster on the right is meant to encourage pilots to fly electronic warfare ships.

they literally became entangled and started bouncing off one another. With the battle spiralling, Manfred Sideous found an opening to align Pandemic Legion's fleet and get them out of this rapidly deteriorating situation and safely off the field. But even as he warped away he could see two CFC heavy interdictors streaking toward TEST's carrier wing. His heart sank, knowing his allies would now be stuck, unable to retreat.

Over the course of the next hour the battle at Z9PP-H escalated wildly, and things started to look precarious for TEST. More than 300 TEST battleships and 30 of its 100 capital ships were now wrecks. TEST knew it had to muster some kind of response.

"With carriers dying at a steady rate, TEST decided to bring in more subcapitals in an attempt to either save their carriers, or at least slow their deaths," wrote CFC Fleet Commander Kcolor, the day afterward. "More Rokhs (DPS ships) hit field, along with a small anti-support fleet. [...] An attempted "doomportal" (stealth bombers bridging into the middle of the fleets and launching bombs one squad at a time) by [Pandemic Legion] was foiled by the large amount of anti-bombers the CFC had on field."

As the CFC's reinforcements streamed into the battle, 4000+ people gathered on Twitch [big numbers for

that era] to watch what was shaping up to be "the next Battle of Asakai." As the CFC closed in on TEST's capital fleet some had even said it might be the killing blow in the Fountain War. A familiar process of escalation took place, as both sides continued to ping their networks for reinforcements. The in-system population soon topped 2000, and showed no sign of stopping.

CCP Games saw this happening and wanted to try to smooth out performance of the battle. The server administrators opted to reinforce the server node that ran Z9PP-H. Once a battle is underway there is nothing that CCP can do to add computing power to the server node that is simulating it, because any change would require resetting the node and disconnecting everyone inside it. The only thing that can be done is to reduce the number of other things that the server also has to do so it can focus on the battle as its singular task. Since any given node operates as many as 7-10 other star systems on a normal day, the other systems nearby Z9PP-H could be remapped to a different node to ease the load on the main battle. Those systems would experience a brief disconnect, but this was fine since the major action taking place was confined to Z9PP-H. This would help smooth out a small amount of the lag in the system and hopefully allow the players to have a functional battle.

"CFC had the entire TEST Capital fleet at their mercy," wrote Trii Seo. "The node was under heavy strain and CCP chose to remap other systems away from that node. However, when typing in the whitelist of systems that would not be remapped, the [server operator] slipped and typed 'Q9PP-H' rather than 'Z9PP-H'."

Two thousand and five hundred *EVE Online* players from around the world blinked bewilderingly at a black screen with the same message:

"Connection Lost: Node is being remaped [*sic*]."

The uproar was swift and vigorous as thousands of angry players clicked "OK" in the error window then slammed "submit" on their livid forum posts.

It was actually worse than anyone yet knew. If this had been a simple service interruption, then the TEST and CFC fleets would have stayed in space as their pilots were knocked offline (as had happened during other high-profile server crashes.) Things would have been chaotic, but the battle would continue and TEST's fleet would most likely still be destroyed.

However, the remapping process was more like the process that occurs at downtime, which completely removes all ships from space even if they were engaged in combat. So TEST's fleet—which was nearing destruction—was removed from space entirely and because of that its pilots didn't have to log back in. The ships were technically still there—or they would be the next time the character was logged in—but that needn't be anytime soon.

With thousands of missiles and lasers streaming toward the TEST capital fleet, its ships were saved and removed from space in what became known as "The Miracle at Z9PP-H."

WITH RENEWED ZEAL

TEST's fleet was saved for the moment, but this incident ignited a long-simmering fire in the CFC, who saw this event as yet another incident of CCP malfeasance. The narrative began to develop that CCP was once again conspiring against the Goons, and that they had pulled the plug on the node on purpose to save TEST.

The CFC experienced a wave of its own as fleets were suddenly overstaffed and full of members ready to once again answer the call to defeat what they saw as some rogue CCP developer's attempts to intervene in the sandbox of *EVE Online*.

The military situation looked bad, but that's not the only issue TEST was facing as the CFC's rejuvenated fleet bore down on them. Even with its hide temporarily saved in Z9PP-H, things were going poorly inside TEST Alliance Please Ignore.

A week later, BoodaBooda—back from the Testival—held another State of the Alliance meeting in which he explained TEST's finances were in dire straits after the mass loss of capital ships at Z9PP-H and continued Confederation of xXPIZZAXx attacks on its moons. He broke the news that if TEST was going to survive it needed a mass of personal donations from the pilots.

The members were generous and the donations flowed in and helped stabilize the stressed social group, but it was a less than proud moment for an alliance that once saw itself as the heir apparent to *EVE* itself.

"If we're going to continue enjoying the hell out of this war, everyone in TEST has to want it," BoodaBooda said in the address. "We need to be intensely focused. We need to be as strong as possible. We ALL need to give this war 100%."

The tone of his speeches had shifted considerably from his original "WE'RE GOING TO KILL THESE NERDS" style, and indicated that things were going even worse than most people knew.

Among TEST's commanding officers, things weren't going much better. It seems that a dispute broke out between TEST Military Director Beffah, who was experiencing strife in her personal life, and 20-year-old CEO BoodaBooda, who couldn't understand her lack of commitment to the video game at the most important time in the alliance's history.

Beffah eventually defected mostly-amicably to Pandemic Legion, citing a "cult of personality" in TEST centered around BoodaBooda. A player named Ingen Kerr was appointed to replace her. Ingen Kerr looked at the situation as he came in and decided that no credible defense of Fountain could be made with TEST's supercapital allies gone, the walls breached, much of the capital fleet still logged out in Z9PP-H, and CFC fleet commanders setting their sights on 6VDT-H. Kerr and BoodaBooda instead decided on a climactic last stand as a fitting end to TEST's days in Fountain. Some were surprised by the announcement that 6VDT-H was to be their Alamo.

"6VDT-H is not anywhere near the level of strategic importance it once held," wrote CFC fleet commander Vily. "Its use as a transit hub is mostly deprecated, and it no longer represents an active military base for more than a fraction of TEST's forces. However, the morning after the initial attack, we awoke to a surprise: All TEST operations for the following two days had been canceled. All forces were to prepare for a defence of 6VDT-H on Sunday, the 28th of July. This was where they would hold the line. This was where they would end our advance."

In truth, by this point TEST knew that the defense of Fountain was lost, and now the only thing TEST leadership wanted was to go out with a bang. They wanted to create a historic send-off to the Fountain War and the pilots who had sacrificed most of their summer fighting it. The chyron above the main station in 6VDT-H read, "BoodaBoodagrad."

One last charge for 6VDT-H, where TEST's journey in nullsec began. ●

THE BATTLE OF 6VDT-H

"As we continue to clear the battlefield—killing everything but the most difficult to track of frigates—word comes in of... something odd in *EVE*. It seems that TEST Alliance plans to make a charge—a last stand—to come back in against overwhelming force and make a statement. In *EVE*, this is not done. You do not fight when there is no hope."

— Vily, CFC Fleet Commander

The Battle of 6VDT-H was the largest battle in the history of *EVE Online*, and led directly to the collapse of one of the largest player organizations in the history of the game. TEST Alliance would subsequently lose its holdings in the Fountain region, lose funding to support such a large group of players and quickly fall apart.

It was to be the culmination of the ClusterFuck Coalition's Odyssey campaign: the wrenching poverty that The Mittani sought to avoid would now be TEST's to endure. However, TEST's performance in the battle would become the stuff of legend, and it helped establish the alliance as an institution in the *EVE* community that rose again in later years. This is the story of the doomed last charge of TEST Alliance Please Ignore.

DOOMED

On the day of the battle, July 28, 2013, the two goliaths began rumbling to life. Both TEST and the CFC started pushing pilots to get online and to organize into fleets. The number of participants in the battle would eventually grow to more than 4000 players, surpassing the Battle of Uemon as the largest battle in the history of online video gaming up to that point.

The logistics necessary to pull off a battle on this scale are staggering. First of all, you can't just snap together an army of thousands in minutes or hours. In order to get 2000 players to show up for a battle you have to build a

Above: The Gallente Administrative Outpost in 6VDT-H, which was often the main hub of the Fountain region in 2013.

message that will echo throughout the digital universe. Many players will beg off due to work, obligations, or personal disinterest. To achieve 2000 pilots, then, you'd better get your message out to tens of thousands of pilots. Building that kind of a message takes time, craft, and a platform (its not a coincidence that both of the core groups doing battle were backed by social networks larger than themselves.) To get that many people out to the battle requires a story that clearly defines the purpose of the battle, what would be gained in victory, what would be lost in defeat, and what constitutes victory or defeat. All players need to understand these things in order to achieve a critical mass that moves the popular body. A fleet like this is an entire community of people given singular animus within a virtual environment, and the story must express the dire importance of the community coming together to take action.

This is why major battles tend to happen over either clearly important strategic infrastructure or symbolic community hubs. To rally the average *EVE* player, who only logs in every few days the stakes must be abundantly clear and historically compelling. But *EVE* is a complex and interconnected social system, and so when an alliance attempts to rile up its members and get them excited to participate in a war, they will also—ironically—end up exciting their enemies and driving up their enemy's participation.

The stakes of a battle need not be victory or defeat in order to rally the players. TEST leadership could no longer promise victories, so Ingen Kerr and BoodaBooda promised instead only that the battle would write TEST into history. And though they have mocked the phrase for more than a decade, there's nothing *EVE Online* players love more than being able to say, "I was there." To get this

215

message out, TEST Alliance—of course—used Reddit, the social network from which the organization had sprung. Having the front page of the r/eve subreddit covered in talk of war and of a great last stand also made it easier for the CFC to gather an armada of its own (though they publicly derided it at every opportunity the CFC's pilots read r/eve too. Strictly for the memes, of course.)

All of this to say, both sides could hear each other's war drums, and the number of people attracted by the commotion was historic. ClusterFuck Coalition Fleet Commander Vily later said that on the day of the battle, before a single call-to-arms had been posted, more than 2300 pilots were already in the CFC staging base waiting to be given orders.

Two-thousand three hundred. The CFC forces alone were more than 10 times the number of ships that fought for both sides in the climactic opening battle of the Great Northern War in 2004. And the battle hadn't even started yet.

Conservatively we could probably say that 25% of the ships were alts, meaning a fleet of 2300 pilots is probably around 1600 individuals. The coordination it takes to move a group of 1600 people is not to be underestimated. It takes clear messaging, and an ironclad grasp of gameplay mechanics. A fleet commander coordinating this many people needs to understand that everything is going to take ten times longer than it should take because 1600 humans are a clumsy bunch. Some of them won't even hear the Fleet Commander when an order is issued. Maybe that pilot was in the bathroom, or talking to someone else, or playing a different game in another window. Fleet commanders have to be realistic and simplistic in their expectations, and patient while they wait for their orders to be completed by 1600 mammals who have no herding instinct. The maddening, menial reality of trying to accomplish simple tasks with hundreds or thousands of people like this is probably also a big part of why some *EVE Online* Fleet Commanders have been known to descend into screaming rage fits.

But not Lazarus Telraven—one of the CFC's lead fleet commanders—who was no stranger to high-stress gaming. He was an esports champion in a first-person shooter called War Rock, a free-to-play South Korean game in the style of Counter-Strike. When time dilation slowed *EVE* down, Laz would turn to his second monitor and play World of Warcraft 3v3 arena matches to pass the time between volleys.

Patience was especially necessary because the CFC's 2300 ships were more than enough by themselves to bring the server to a full 10% time dilation. That meant time was moving at 1/10 it's usual speed but only within this one star system. And even then it didn't clear up lag completely, it just helped a lot. Combined with additional latency things could move very slowly on the battlefield, often leaving players with hours of downtime between major fleet movements. Keeping people focused and attentive throughout a battle that was sure to take all day was one of the trickier parts of commanding a fleet of video gamers.

"Moving such a mass of forces is an extremely unique aspect of *EVE Online*; moving a single 256-man fleet can be a challenge at times, and moving eight of them at once is even more so," wrote Vily in his retrospective. "Time dilation usually kicks in any time you see 300 or more pilots moving as a group, so moving 1000+ makes it a certainty. As such, the CFC formed almost two hours in advance of the reinforced timer exiting invulnerable mode. It is not fun, but it is absolutely necessary if you plan to achieve system control first, giving you a significant tactical advantage."

Vily and the CFC sky command carved that 2300 into smaller, more easily controlled fleets. Their main damage would come from 7 full fleets of Baltec Megathrons of 256 ships each (the max fleet size) led by Fleet Commanders Vily, Mister Vee, Lazarus Telraven, Reagalan, Cor Six, Intergalaktor, and Ironwulf. Two fleets of bombers were formed under Kcolor and DaBigRedBoat. A fleet led by the player David Cedarbridge was dedicated to electronic warfare (warp scrambling, mana draining, that sort of thing.) Imperian—who had years before led the Northern Coalition's defense against the Drone Region Federation—led a group of ArmorHACs, adapted from Pandemic Legion's original killer blueprint, in what Vily called a "shark hunter role." Meanwhile, Fleet Commander Theadj commanded a wing of carriers and dreadnoughts, waiting in an adjacent system for the signal to move in. Everyone had a job to do.

"Our job was to scan down disconnected players—preferably command ships and boosters—and kill the enemy ones to ensure they did not rejoin the fight after reconnecting," wrote Trii Seo who self-identified as a "cloaky bugger."

The full CFC skycommand team took control of nearly a dozen different fleets with different jobs to do in the taking of 6VDT-H.

Above: A small frigate watches from the distance as a fleet surrounds and destroys an iHub.

As the first fleet in the system, the CFC had its choice as to how it wanted to set up its fleet in preparation for the battle. The CFC Sky Command was getting reports that TEST had formed more than 1500 pilots, a historic fleet unto itself but well short of the CFC's 2300. With CFC holding numerical superiority, the fleet commanders reasoned TEST might use bomber fleets to try to kill masses of CFC ships and level the odds. To pre-emptively counter against this possibility, the CFC ordered all seven of its Baltec Megathron fleets—1600+ players—to orbit the main station at a distance of 80 km and spread themselves as thinly as possible.

The result was a spherical cloud of ships encircling the tall outpost which looked like a skyscraper above the blue atmosphere of its moon.

In two hours, the reinforcement timer would elapse, and the station would be conquerable. A pilot for the CFC named Wilhelm Arcturus remembered the day in a detailed retrospective:

"Sunday morning I rose late. The [operation] was set to begin at 11:30am local time for me, so I rolled out of bed a little after 9am and jumped in the shower. After getting dressed and having the "breakfast of champions"

(cold pizza and a coke) I got myself logged into my computer and into the various comms channels with well over 90 minutes left before the form up time. I was greeted by calls to stock food around my desk, to say farewell to my family for the day, and to get online and in a ship and undocked ASAP to avoid the rush.

When I got in-game in 4-EP there were already 1,300 people in the system and that number was growing quickly, even as early bomber fleets assembled and jumped out. I got slowly undocked and joined the mass of Megathrons at the staging POS where we attempted to form a conga line. It was still nearly 45 minutes before the official operation start time when, faced with 2,000 pilots in the staging system and time dilation [moving at 10% speed,] the decision was made to form fleets. [...]

One by one the fleets were bridged out through the tidi. We went by the numbers, so the sixth fleet went last while command was attempting to form

Above: An actual screenshot from the hours before the Battle of 6VDT-H in which the CFC Baltecfleet can be seen orbiting the TEST station.

up additional fleets for people still waiting. Fortunately for us, getting to 6VDT-H involved a single bridge from 4-EP. [...]

We moved to the station where one of the recurring pictures of the battle was [seven] fleets orbiting the station. The view was compared to ants swarming. I preferred to think of it as bees around a potential new hive."

— Wilhelm Arcturus, CFC Pilot

In their thousands, the Megathrons silently proceeded in slow, time-dilated automatic orbits, waiting to see if TEST would be brave enough to show up and fight for its spiritual home.

As the CFC pilots got into position around TEST's former headquarters its fleet commanders prepared themselves for what they believed could be a long day of battle.

"Most of our pilots took advantage of this time to get lunch, say goodbye to their families, and prepare for the worst while expecting the best. At the same time, N3ST [editor: a common portmanteau of TEST and N3] forces slowly went through the process of moving their forces into a position to engage. This can be arduous, and it was obvious that they were playing against the clock to organize themselves properly to arrive in time to contest the system.

At approximately 5 minutes left of the station timer, hostile cynosural fields were spotted in-system and the enemy [N3ST] forces took Titan bridges into system. It's hard to describe the sense of anticipation one feels when preparing for something of this grand scale, and certainly the pulse of myself and many other of the fleet commanders rose. The fight was coming; it was happening."

Vily, lead CFC Fleet Commander
July 28, 2013

The N3ST mega-fleet arrived in 6VDT-H amid seven thunderous spirals of pink electricity. The electric spirals were the sparks of cynosural fields channeled by Titans in nearby systems that teleported the N3ST pilots into the system by the hundreds. Each fleet was led by a Damnation-class logistics ship flown by the fleet commander that was armored and buffed in every way possible. N3ST was doing everything it could to keep its chain of command from falling apart once the melee began. It took about 20 minutes for the full force of N3ST's fleet to arrive in the system.

Once inside 6VDT-H, the battle between the two fleets began before N3ST got much of a chance to prepare; DaBigRedBoat's bombing fleet was already slowly streaking across the sky at 10% speed on a direct course for the center of the N3ST fleet.

TEST Fleet Commanders had plenty of time to think about their next choice. DaBigRedBoat's bombers creeped closer in ever slower motion as lag—on top of time dilation—began to set in. TEST Military Director Ingen Kerr had two choices: 1) absorb the brunt of the bombing and lose hundreds of ships, but hold position, or 2) warp the fleet out of DaBigRedBoat's path and toward the only available location: the Outpost swarming with 1600 CFC megathrons.

Ingen Kerr chose the latter, and his ships warped by the hundreds toward the outpost where the battle would finally truly commence. Through crushing lag, the N3ST Prophecy fleets launched their drone bays and assigned control of them to the lead fleet commander, allowing that one single person to direct exactly where the N3ST fleet's damage would come down with perfect precision.

"The first fleet to land was a segment of the [N3ST] electronic warfare group avoiding the bomb run, but it would be followed in such close succession by Prophecy and [ArmorHAC] groups that it could have hardly been more than a minute between warp-ins," wrote Vily. "Now we were truly engaged. Fire was exchanged, and the massive groups of railguns carried by the Megathron fleets spun into action."

CFC pilot Wilhelm Arcturus offered their recollection: "TEST landed on us and the fleets were intermixed in that odd way that happens in very large fights. At times I found myself flying between a pair of [NCdot] Dominixes or through a cloud of TEST logistics [ships.] A Nulli Secunda logistics Bantam [a small repair ship] appeared to orbit me at one point."

Vily noted that the very first of his Megathron salvos targeted N3ST Fleet Commanders, a tactic he himself seemed to think was underhanded and yet unavoidably necessary since N3ST would certainly try to do the same

to the CFC; and they did. Leaders of both sides gave orders to target the other fleet's leaders.

In a short time it became obvious that the N3ST fleet was having trouble breaking through the heavy armor of the CFC command ships, which heroically faced down the near-simultaneous fire of a thousand N3ST drones. The CFC's 1500 megathrons were having no such problems. One-by-one they carved through TEST's most critical ships and command structure. As the ships burst, CFC pilots were ordered not to target the escape pods so TEST would have to fly the pods back home manually through the aching slow of time dilation.

"Meanwhile, we shot," wrote Arcturus in their retrospective. "Targets were called and we struggled to get our guns to fire on them. [...] I had to change my overview a number of times. First I had to exclude [escape] capsules, which began to litter the field, and which we were told not to shoot. Let them walk home was the plan."

Hours went by as the laser-crossed tangle of ships destroyed each other near the massive starbase. Warp disruption bubbles created a hazy blue hue to the battlefield, coursing with layers upon layers of electricity. By the end of the first few hours, hundreds of ships had been erased by the sheer weight of firepower in effect. As ships were blown off the grid, the servers began to catch up and the lag cleared, allowing the fleets to operate somewhat normally, albeit at 10% speed.

"Merely being in 10% tidi felt liberating," wrote Arcturus. "Simple things, like your guns activating on the first try or targets actually locking in the time indicated were like a breath of fresh air. Running at one tenth normal speed is a doddle, if only the client will actually respond to your actions."

While the average pilot calmly orbited the outpost, the hardest job fell to CFC tech administrator Solo Drakban, who had to keep clear communication lines up between 2300 people. The CFC's historic fleet would be next to worthless if the Mumble server went down.

"Solo Drakban was no doubt fretting over the comms infrastructure," one pilot wrote. "In the past, big operations like this have brought down our voice comms, requiring the server to be re-provisioned to accommodate the load, a process which requires every pilot to go change their Mumble configuration. Since, at the best of times, maybe 4 out of 5 people actually hear (but not necessarily comprehend) instructions coming over comms, that sort of thing becomes a major undertaking on its own.

Above and Below: More angles on the historic fight in 6VDT-H, at the time the largest battle in video game history, and the final battle in the defense of the TEST-owned Fountain region.

"So we were told to just stay quiet and use voice comms as little as possible, lest we bring the whole house down," he continued. "This actually worked, at least on our channel, which was surprising. Goons and allies are a talkative bunch and, when left with little to do, will begin to chat or argue about whatever happens to come up. With the warning in place, for long stretches it was like being in the "no chatter" channel, where you can only hear the fleet command personnel speaking."

Those unusually quiet command channels were relatively calm while orders were going out to thousands of pilots about who to destroy next. The historic CFC Baltec Megathron fleet sent thousands of railgun blasts toward whichever unfortunate target was designated next, and even N3ST's heavily-armored command ships evaporated in moments, as did its command structure.

Vily's report suggests that the N3ST ArmorHAC fleet lost multiple fleet commanders to combined CFC Megathron strikes, and the rest of the fleet was rendered utterly ineffectual for two whole hours as the fleet attempted to get back online and operating under a coherent command structure.

But the most devastating impact was had by the bombing fleets run by Kcolor and DaBigRedBoat. More than 25% of the 2900 ships that would be destroyed in this battle were killed by their bombs. Their fleets of specially outfitted bombers swooped in to exploit tightly-packed groups of uncoordinated N3ST ships which, deprived of effective leadership, had no chance to react in the crushing lag and time dilation. Some of these bombing runs destroyed up to 70 N3ST players at once. The bombs were not without collateral damage, however—an additional 15-20 of the CFC's own ships were usually caught in the bomb blasts as well. Unfortunate, sure, but a small price to pay for the havok the bombers wreaked on the enemy.

Meanwhile, the CFC ArmorHAC fleet under Imperian maneuvered toward the only stargate TEST could use to exit the system, destroying retreating fleet remnants and deserters.

THE AUDIENCE

All the while, several unaffiliated pilots sat on the outer reaches of the system in cloaked ships, passing video of the engagement to the wider community through Twitch streams. In addition to the more than 4000 players in the battle, another 10,000 watched live on Twitch. Streamers Mad Ani and Daopa led the effort to get word out to the rest of the world about the largest battle in gaming history. The CFC's news service "Mittani Media" also operated its own stream and issued updates about the battle in progress.

For the 10,000 who were watching, the battle was an incomprehensible mess. The screen often seemed frozen in time while streamers valiantly tried to explain that the tangle of brackets and tiny ships scarcely moving on-screen were actually two fleets from opposing factions within the game world. One side were the Redditors of TEST Alliance, and their allies N3/PL. The other side were the Goons of Something Awful—a necessary simplification. They explained that the fate of the Fountain region and its valuable crop of moon mining operations lay in the balance. They explained that what was happening here was something of a tragedy. That a long time ago these two communities were friends who had a falling out over politics and ambition. Few knew at the time that TEST's former leader Montolio had coveted exactly such a battle, and his provocations may have played a critical role in eventually causing it.

That 10,000 people stuck around for the stream was remarkable given that none of the actual action described in this chapter was discernible from that vantage point. Mostly, the stream operators tried like hell to keep a live video feed operational, because often their *EVE* client would crash trying to perform actions as simple as adjusting the camera to give the audience a better view. The 8-10 hours of work these streamers put in to get these images out to the world created a rare window into this crucial event inside an often hopelessly opaque universe. For one of the first times in *EVE*, the average person could get a look at major events happening as they unfolded. Even if it looked kinda boring from this vantage point.

"The actual battle was horrible," wrote a commenter on an *EVE* website article about the battle at 6VDT-H. "I actually managed to clean the house in between when I deployed drones, and when they shot something."

Another pilot at the battle named Jonathan Stripes wrote: "It got to the point, after 3 hours, that my *EVE* client's clock was no longer running, and I got this error, stating that the server was not even receiving commands from my computer any more. Many laughs were had on coalition comms about the errors, and the FC just saying "fuck it, if you can target it, shoot it" made for one of the most fun gaming experiences I have ever had. For this fight, in which more than 4000 pilots were in system at the same time, CCP set the system up on Jita's dedicated supercomputer."

However, the server began to catch up as more and more TEST pilots were cleared from the field, and the remnants of the fleet at last beat a retreat toward the star gate that led to safety in Delve.

THE LAST CHARGE OF TEST

As TEST reached the escape stargate it encountered Imperian's "shark hunter" ArmorHAC fleet, and began losing ships immediately. TEST command began to realize that the price for extracting from this system would be grueling and difficult. It would take hours, and Imperian was likely to take as many as 30% of them down before all of the fleets managed to get out of 6VDT-H. With plenty of time to talk it over in 10% time dilation, TEST command decided how TEST would end it's final battle.

"As we continued to clear the battlefield, killing everything but the most difficult to track frigates, word came in of something odd," wrote the CFC's Vily in his retrospective.

Ingen Kerr considered TEST's actual goals for the battle, and remembered that he had not actually come here to win, but to give this great adventure with TEST the finale it deserved. He ordered the fleet to stop, and told his fleet commanders to turn about. He told them to charge into the CFC formation, and target the most expensive dreadnought they could lock.

"It seemed that TEST Alliance planned to make a charge – a last stand – to come back in against overwhelming force and make a statement," Vily wrote. "In *EVE*, this is never done; you do not fight when there is no hope. You retreat and save your ships for another day, another chance. But in they came."

The final charge wasn't about turning the battle around and saving 6VDT-H or Fountain. It was about saving TEST Alliance itself from succumbing to the same God-Save-The-Ships ethos that had dissolved so many groups before it.

And so TEST mounted a final charge—not to win 6VDT-H—but to save its collective soul. Hundreds upon hundreds of TEST ships streamed toward the outpost at full speed back into the unbroken CFC defense. "Full speed," however, was 25% what it usually was, and the CFC had several minutes to discuss how to perfectly dismantle the incoming TEST remnants. The TEST ships turned from red to grey by the dozens on the overviews of CFC pilots as they were coolly picked apart by CFC

fleet commanders who couldn't help but feel a pang of respect for the unprecedented sacrifice. Every single one of TEST's ships would be executed in just a few minutes even as their many lasers converged on the hull of a single CFC dreadnought. They managed to destroy their symbolic target just moments before the last of their fleet was cleared from the field. The bubble of Megathrons now slowly orbited in silence beside a great sea of debris.

Today the explosive final charge at 6VDT-H is remembered as the climax of the Fountain War, a way for TEST to go out on its own terms. It was a statement, as loud and as clear as the 27 Doomsdays that marked the end of the Great War; a declaration that TEST would not allow its identity or decisions to be dictated by an outside force. If sacrificing the entire fleet to kill just one final Goon was stupid, then let history carve that stupidity into stone and call it the legacy of TEST Alliance Please Ignore.

TEST leader BoodaBooda gathered the remnants of the alliance as they logged out of the battle, and delivered a speech to hundreds of pilots on Mumble.

```
"Today we fought in 6VDT. We didn't fight
to save the station, we didn't fight to
win an ISK war, and we didn't fight to turn
the war around.[...] We fought because TEST
stands together in the face of certain,
guaranteed death. I wanted to give Fountain
one HELL of a sendoff, and it's incredibly
appropriate that it took place in our home
system of 2+ years.

     This war has taught us many things -
We've learned what we can do at our best,
and we've learned what we need to do to
properly support our FCs, leadership, and
membership.

     We made an absolutely fantastic showing
today. We fielded more members than any
single alliance has ever seen - we even had
twice what [Goons] alone brought - and we
set a record for the largest fight EVE has
ever seen by a few hundred pilots, maybe
even a thousand. [...]

     So with all that said, we've been losing
a defensive war for about two weeks now.
As soon as our big strong capfleet-toting
allies had to run home, the CFC pulled out
```

all the stops, and TEST was unable to keep up on our own. They practically rolled through Fountain unopposed.

This isn't the fault of our members. It isn't our FCs' faults, or our [Military Director's] fault, or [logistics,] or recon, or our corp[oration] CEOs… This war has been an immense team effort, and one of the most significant things we've learned is that our team needs some improvement before we can function together well enough to pull off a large scale war like this. […]

We need time to do the work we should've done 3 months ago. 3 years ago. We have to clean up our leadership groups and structure in order to handle something as significant as a full-on coalition-level war—or something as simple as record-breaking activity numbers.

The CFC will almost definitely attack Delve, either in a continuation of their full-scale invasion, or with prodding tests. It will give us an opportunity to put some strain on a rebuilt leadership structure and see how we function. I will be constantly reevaluating our position the entire time, but for now that's the plan. [...]

I can only hope that you don't feel I've thrown away your donations to the alliance. Ultimately, the sov is here so we have something to fight about, and I felt it best to spend our cash in the bloodiest, largest, most amazing fight the *EVE* community or even the video game community at-large has ever seen.

So fellow TESTies, follow me back to Delve, and let's set out on a journey to make TEST the alliance this community deserves."

— BoodaBooda, Executor, Test Alliance Please Ignore

July 29, 2013

"The war was over, and the CFC flags rose above the smoke of 6VDT-H," wrote CFC pilot Trii Seo. "What once was a hostile ground [was now] a bustling hive of regrouping and celebrating CFC forces. It's hard to describe the feeling exactly–months of warmongering and smacktalk, and the smacktalker was running towards [Delve,] tail between his legs. If I could stick a flag out of the window of my Nidhoggur [carrier] as I flew, I would totally have done that. Played Chariots of Fire maybe."

Within days of TEST's retreat into Delve, BoodaBooda got word that the CFC would not be stopping with Fountain, and that attacks on Delve were being planned for the next few days. The situation for TEST rapidly deteriorated. With the situation as it was, he soon stepped down.

"I'll admit that constant betrayal, daily ultimatums from corp leaderships, and utter lack of motivation across the board sapped out any energy I had left for TEST," he wrote in a goodbye letter announcing his resignation. "Ultimately, I stepped down because I'm exhausted with leading, with politics, with diplomacy, and with relying on people I cannot trust to stay and cannot trust to do the right thing.

"So I did something that I don't think I've ever done in *EVE*," he wrote. "I did something that I stopped doing in my personal life many years ago. I quit."

The difference in BoodaBooda's tone from the beginning of the war is stark. One of the most tragic parts of *EVE* that people rarely talk about is that not every player gets to "win *EVE*" and go out on their own terms. If it's possible to win *EVE* it's also possible to lose it. Both, it seems, depend on what condition *EVE* leaves your spirit in.

THE GREAT PURGENING

BoodaBooda's departure only lasted about six months until he returned in an attempt to help TEST grapple with the next era in its history. Now he wanted to finally fulfill his vision for TEST, and start again from the ground up. On August 10, 2013, he and the TEST team went to the member roster and removed anybody who hadn't logged in since before the war. They streamed the whole thing live in what turned into a sort of celebration that renewed the pact with the community. They called it the "Great Purgening."

"It's time to finally start the next leg of our adventure through *EVE*, with a largely cleaned-out alliance rebuilding around us as we rebuild ourselves in kind," he wrote. "Let's see what this game still has to offer. But more importantly, let's tackle it together." ●

THE HALLOWEEN WAR

"The RUS/CFC fleets either forgot the lessons of old or were simply overcome by greed. On the brink of Supercapital Armageddon, all hell broke loose."
— Elise Randolph, Pandemic Legion

The ClusterFuck empire now mirrored eerily the dominion of the former Band of Brothers at its peak save for the fact it was a bit larger and more stable. From the far north in the Tenal region to the newly conquered deep south of Period Basis, the territories controlled by the ClusterFuck in 2013 were almost exactly those owned by Band of Brothers at its brief apex in 2008.

For only the third time in New Eden's history, half the nullsec territories were controlled by a single organization of players united under a single ruler. Many of the denizens of *EVE* jeered that the leaders of the ClusterFuck—Goonswarm Federation—had completed a sort of despotic life cycle, beginning as a rebellious force fighting the entrenched powers, only to become a mirror of those very entrenched powers once that same power was in their hands.

"One cannot help but feel that the [Goonswarm Federation] has evolved from its roots by quite a bit," wrote Seraph IX Basarab, who often wrote from a virulently anti-CFC perspective. "From their days of throwing swarms of Rifters at the "evil" BoB Empire, hanging out with Red Alliance and being a goofy every-man's alliance, they have become quite arguably, THE most powerful entity within *EVE*. [...] Where once they swarmed around the tall walls that BoB had put up, mocking them for their decadence of rental programs and elitism, the GSF now find themselves within their own walls, with their own decadent rental program and it seems also engaging in their own level of elitism within the CFC. Those, it seems, that aspire to have an identity outside of the specific GSF mindset are deemed 'heretics, mutants, and unclean.'"

One TEST Alliance pilot sneered on the forums, "The Mittani had completed his transformation into SirMolle, a Jabba the Hut of idiotic proportions."

THE COUNT OF MITTANI CRISTO

As the fires of the Fountain War were being stamped out, the stage was already being set for the next set of engagements which would consume nullsec. The rapidity was no coincidence. Many had seen the Fountain War as a sort of proxy war against the loosely aligned force of NCdot, Nulli Secunda (the core of N3,) and Pandemic Legion.

Even now the next stage of that war had begun through a sequence of events that started small and escalated. The mercenary group Black Legion's sister organization—Confederation of xXPIZZAXx—was employed by the CFC to hit some shipyards in the southeast that it had learned were building NCdot's future supercapitals. When NCdot fought off the attack, xXPIZZAXx called up Black Legion to add its formidable capital fleet to the assault. Things soon spiraled even further, as N3/PL arrived to aid the defense and Black Legion had to call in the CFC supercapital fleet to critical engagements.

Opposite: An entire fleet of Titans packed so thickly they nearly obscure the stargate they are gathered around.

One night, Pandemic Legion fleet commander Makalu Zarya left a Titan out of position as bait. When Black Legion attempted to destroy it, the trap was triggered and escalated wildly leading to a brawl between thousands. As October 2013 came to a close, it was already becoming clear that a new bloc-level war would soon begin. By now few were even surprised, as it was clear that in some respects *EVE* was in the midst of a forever war. The individuals that made up these alliances could not be destroyed, and neither could their very real animosity for one another which was compounding through the years.

What few knew at the time was that the ClusterFuck Coalition was beginning to fray. What's more, the CFC's Stainwagon allies in the south were embroiled in near constant in-fighting and on the brink of collapse. Meanwhile, the coalition's relationship with its Black Legion mercenaries was fraught with distrust; Lucia Denniard of Confederation of xXPIZZAXx said The Mittani and Sion Kumitomo repeatedly made accusations that Lucia and frequent collaborator Elo Knight of Black Legion were secretly plotting against the CFC, even as they worked with them. It was an accusation Lucia Denniard acknowledged and denied in the same breath.

"People knew I had a large spy network which made them distrust me and see ghosts everywhere especially when I was teamed up with Elo [Knight] who occasionally backstabs allies openly because it's more fun," Lucia Denniard told me.

The CFC gathered its shaky allies for a follow-up invasion to capitalize on its success in the Fountain War. It was anyone's guess whose rickety coalition would survive the coming contest. With TEST out of the picture for now, the CFC's geopolitical adversaries were now outgunned and outnumbered, and yet they had faith in what they believed was a superior understanding of the game and superior amount of skillpoints (since their pilots had generally been subscribed to the game longer. Those two factors combined allowed them to field entire fleets of Slowcat carriers, which nearly drove the CFC to madness.

The CFC had one clear advantage, however. The alliances of the ClusterFuck Coalition owned a majority of all the Titans in New Eden. While CCP Games had originally envisioned only a few dozen ever being constructed over *EVE*'s entire lifespan,

there now existed a stockpile of approximately 1500 across all of *EVE's* alliances. New Eden's engineers were pumping out new Titans at a rate of 60-80 new ships every quarter; ten times faster than they were being destroyed. If an all-out battle was to break out, the N3/PL forces would have to destroy three CFC Titans for every two they lost—a tricky proposition when you're outnumbered and overpowered from the outset. And to make matters worse, the odds were getting more severe all the time, because the CFC owned more profitable territory. The only way to survive was for N3/PL fleet commanders to develop new strategies that allowed them to fight outnumbered.

Nearing the conclusion of 2013, the question that hung over nullsec was whether the CFC would eventually fall apart, or would The Mittani be able to find a way to strike a killing blow against his only serious coalition-level opposition bloc. Would he become the emperor of *EVE*? Or would he be beheaded by the very same Sword of Damocles he had warned SirMolle about six years before?

ALL HALLOWS' EVE

"In the first week of the war, [the CFC] attacked our home constellation," said Nulli Secunda's ProGodLegend. "They tried the headshot. They got really close once, and we stopped them. They got really close a second time, and we stopped them again. At that point they said 'OK this isn't going to work, and they went for a longer war of attrition. So after that first week they stopped assaulting Nulli's home constellation. And Goons started bringing more and more ships down, and telling their members to take the war in the south more seriously."

With the two predominant coalitions in a state of war, the other alliances of the star cluster were forced to pick sides. The arrangements were complicated: Solar Fleet and RUS sided with the ClusterFuck Coalition, while what was left of Legion of xXDEATHXx sided with N3/PL out of hatred for Solar Fleet. However, the ClusterFuck Coalition and Legion of xXDEATHXx had managed to stay on relatively friendly terms, and they agreed to a non-aggression pact privately between themselves, with xXDEATHXx reserving its fire for Solar and RUS.

While the alliances of the star cluster picked sides, the ClusterFuck Coalition launched a campaign that it called the Halloween War—so named simply because it was close to Oct 31—which was framed as a punitive campaign against N3 for its role in the Fountain War.

In a CEO update, The Mittani fanned the flames of animosity, writing about N3's use of "laughable mechanics" and announcing the beginning of hostilities on what he sometimes called "the Ostfront," which means "Eastern Front" in German and was the name Nazi Germany used to describe the war with the USSR during World War II. It was not the Goons first overt reference to Nazism. The very name of the corporation he was CEO of was "GoonWaffe," a reference to the Nazi air force. In the following quote, The Mittani will also evoke the imagery of "putting our jackboots to their throats." In 2013 these references often flew under the radar, discarded as shock humor. In more modern times, they require closer examination. A sociological study of Goonswarm's posting history is beyond the scope of this book, but a few details help provide some context.

Goon leaders and others throughout *EVE*'s history have often invoked macabre historical imagery to paint their in-game proclamations in the aesthetic of totalitarianism. Because they are rulers of a cold, dark, and harsh universe they make callbacks to the darkness in our own shared history. At the height of the Great War, Darius JOHNSON referenced the conquistador Hernan Cortez who committed genocide against the Aztec people. The capital station of Goonswarm was named "Mittanigrad," casting his name in parallel with Stalin's. Many alliances repurposed Nazi and Soviet propaganda, and some of the darker memes even referenced modern terrorism. But to many Goons this is all just a silly game, a "sci-fi themed masquerade ball" as The Mittani would later describe it. The Mittani himself, for instance, has expressed progressive personal political views in my interviews with him and on social media. He occasionally even mocks real world dictators who are doing an amateur job at despotism and points out when they are abusing his favorite tactics.

I can say that the references make me personally uncomfortable, but I think the Goons would probably think that was pretty great. Mission accomplished, they would likely say. That is what drew them together in the first place: a belief that their tolerance of risqué and lurid content made them an ideological tribe. That said, it's easy to see how risqué Nazi references can attract and give cover to actual Nazis. Which is a very real problem in *EVE Online*.

While it may not necessarily indicate a fondness for fascism in the real world it most certainly reflects a fondness for authoritarianism in the digital world. In the following speech The Mittani utilized one of the dictator's sharpest tools: victimization.

"They come for us in Syndicate, they come for us in the Great War, they come for us in VFK, they come for us in Fountain, and now they act horrified and surprised when we show up in the East and start putting our jackboots to their throats. They actually have the audacity to be surprised that we would come for them, hungering to inflict the vengeance they so dearly deserve.

Northern Coalitiondot is literally a mishmash of Band of Brothers and Raiden, minus Reikoku—which is sharing fleets with them regularly. Nulli Secunda is no BoB, but they made their bed when they loudly declared in the Fountain War that their entire N3 coalition exists to try to destroy the CFC—our destruction is their raison d'etre. Never forget this. [Editor: This is largely true. He seems to be referring to an alliance update written by ProGodLegend of Nulli Secunda during the Fountain War in which he used the phrase "raison d'etre" and said that the current war is the coalition's reason for being. However since it was a defensive war it's not clear from context whether he meant the CFC's 'destruction' as The Mittani alleges.]

[...] They yet again have another pathetic, indefensible and shameful gameplay mechanic in their hands which they will use to try to destroy our people, just as they always have [editor: he's referring to the Slowcat

carrier.] That is what they do: find
some bullshit mechanic, abuse the hell
out of it, all the while defending it
vehemently and simultaneously trying
to annihilate us.

[...] Our ancient enemies have found
themselves another superweapon [edi-
tor: we'll talk about this soon] and
so we're going to overreact on an absolutely colossal scale and smash them—
and all their friends—into dust and
tears. We will smash them in Immensea,
and put a stop to their dreams of our
ruin before they get out of hand. Once
we have the largest fleet of Dreadnoughts the game has ever seen, we can
give the entitled shitlord demographic
the finger and have a round of carefree
hijinks elsewhere.

Until then, it's war."
— The Mittani, Goonswarm Federation,
Clusterfuck Coalition
Dec 15, 2013

The opening month of the war featured frenetic fighting between two blocs that by this point genuinely hated one another. Conflict after conflict over the years had embittered many of their members against each other. The beef was sometimes personal. In particular, one source told me about an escalating series of conflicts between Manfred Sideous and a CFC counter-intelligence agent who gained a reputation for taking things too far.

"Goons found Manny's Titan spy, and killed the Titan," that source said. A "Titan spy" would refer to a spy that was so valuable it flew a Titan in the enemy supercapital fleet so it could follow its movements. "They called a Titan operation, and then they said on comms, 'Manny we know who your spy is,' and then they killed the Titan."

The grudge escalated even further, however, and Manfred Sideous was doxxed [publishing of a person's personal information on the open internet.]

"Manny had a grudge against Goons that Shadoo did not," the source said. "Shadoo had worked with Goons many times. But Manny did not like Goons, and he still doesn't like Goons to this day."

That source also claimed to have been doxxed by the same CFC spy. "He crossed the line quite a few times. And it ended up pissing a lot of people off. He crossed the line with me once when he burned my spy. Just to give you an idea of how this plays out. One time I had a spy in a Goon fleet. [But they] had my IP number and I wasn't using a VPN. So he found out who my spy was just by comparing the IP numbers. Instead of just kicking my spy—and this was a full Goon fleet, so there's 250 people on comms, I don't know any of these people, they're all internet strangers—instead of kicking my spy and saying gg, he goes on comms and says, '[Source's full name and address] and whose ISP is Cox Communications, we found your spy, bro.' It happened a lot. He doxxed a lot of people. The worst was Manny. No one got doxxed worse than Manny. Manny held that grudge for a long time."

The Halloween War would go on for about three months, and while much of it included frenetic fighting between the two blocs, it was only the last three battles of the war that truly mattered. Of those three, two were so climactic that all the maneuvering and positioning and politicking before hardly amounted to much.

The first battle will fizzle out without much incident. The second will herald imminent doom for The Mittani and the ClusterFuck Coalition. The third—which we'll see in the next chapter—will end this story.

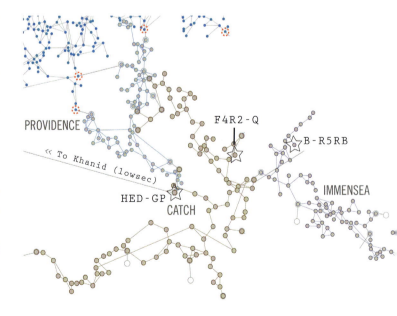

F4R2-Q

Just after the new year, Against ALL Authorities' home system was attacked in a simple harassment operation that cascaded into a major problem. Morale was in the pits inside the alliance, and its own leader was discovered berating the membership for low activity in a forum leak posted on EVENews24.

"Fleets are called, pings are sent, not one fuck was given," wrote the leader of Against ALL Authorities. "Alliance has the same trolls talking that never [get into] fleet. And the rest? Silence. Not a peep. Not a word not a fuck given. [...] Our participation hasn't gotten better, it is worse. Excuses seem to be the Fleet Doctrine of today's [Against ALL Authorities.]"

With membership activity and morale at a new low, Against ALL Authorities accidentally forgot to repair its home starbase and let it fall into its final reinforce cycle. In other words, the home system of the RUS coalition was suddenly vulnerable to conquest. In order to protect their southern RUS ally and keep it from falling apart, the CFC responded with a massive force to safeguard its ally's capital station until server downtime.

"[Against ALL Authorities'] bacon was saved by their allies who proceeded to put a combined 2500 man [capital and subcapital] fleet into F4R2-Q to guard the system from downtime – about eight hours. PL/N3 were simply astonished by this showing and couldn't fight; catastrophe averted for the ever downtrodden [Against ALL Authorities]," wrote Pandemic Legion's Elise Randolph afterward.

Pandemic Legion Southern Commander and former AAA leader Manfred Sideous was disappointed. This was a golden opportunity to gain back some ground in the war, and it slipped through his fingers only because of an utterly miraculous turnout from the CFC membership. However, Sideous resolved to use this information as best he could. Now he knew it was possible to engineer a situation that forced the ClusterFuck combined fleet to fight. So he set to work trying to recreate the circumstances that led to the CFC Sky Command deploying 2500 pilots to Against ALL Authorities' capital. Only this time he'd be ready to exact his revenge upon the Goons.

Sideous selected a system called HED-GP in the region of Catch to set a trap. There he planned to force the CFC fleet into the open and unveil his newest fleet strategy, the superweapon The Mittani plainly feared: the Wrecking Ball.

HED-GP

The system HED-GP was chosen because it has an extreme lack of moons. That star system has only six moons where players can anchor starbases. Sidious slowly took over all six of the moons in the system while leaving it nominally controlled by Against ALL Authorities to avoid raising any alarms. Once all six of the moons were under Pandemic Legion control, then Manfred Sidious led a fleet to HED-GP to attack the system's iHub and bombarded it into its reinforced invulnerable state.

CFC and RUS weren't too concerned about this yet. After all, they could simply anchor a cynosural field jammer in the system and then N3/PL would be unable to bridge in its capital fleet. Without the capital ships, the N3/PL forces would have to fight using only small subcapital ships, which would give the far more numerous CFC/RUS fleet a massive advantage. But when Against ALL Authorities arrived in HED-GP and tried to anchor the cynosural field jammer, they quickly discovered that there were no starbases around any of the star system's moons to anchor it on.

It quickly began to dawn on RUS leaders that Manfred Sideous had been planning this for weeks and that HED-GP was going to be the site of a major attack. On the up side, they were abundantly confident coming out of the non-battle at F4R2-Q, and felt this was an opportunity to shut down N3/PL once and for all. CFC/RUS wasn't planning on just holding the defense this time. They wanted to make Manfred Sideous pay for this. Against ALL Authorities contracted Elo Knight, and Black Legion's imposing 200-ship Dreadnought fleet—the largest such third party force still operating in nullsec—was put on standby in preparation for a major engagement.

"The gauntlet was thrown; HED-GP could only be saved with a similar capital and supercapital showing from RUS/CFC," wrote Pandemic Legion's Elise Randolph. "As happens ever so rarely in *EVE Online*, the stars aligned and a truly awesome [capital ship fight] was on the horizon."

Both sides had days to prepare for the battle at HED-GP, and the forces that arrived for the battle were some of the largest that had ever come to a single fight. The CFC arrived with its historic Dreadnought fleet, 700 strong, flanked by fleets of logistics ships and DPS [damage-per-second] battleships.

Manfred Sidious arrived in HED-GP at the head of "500 Archons, 100 Dreads, 170 Supercarriers, and 70 pivotal Titans" which Elise Randolph speculated was likely the most imposing force ever assembled by N3/PL at this point in its history. This was Manfred Sideous' first opportunity to test the Wrecking Ball strategy, a fleet concept which he hoped would save his whole coalition.

The Wrecking Ball was a complete oddity in *EVE Online* fleet strategy, and it relied on previously unheard of coordination. Most fleet concepts in this era of *EVE* revolved around maximizing the amount of damage a fleet could do in sharp bursts to destroy ships in one shot. These fleets were generally giant balls of ships that either fired in unison to destroy enemies in one volley or used drones which essentially did the same thing.

The Wrecking Ball was the opposite. It revolved around making ships invulnerable. Manfred Sidious used a sphere formation of hundreds of carrier ships to essentially create a bubble of friendly ships. Five hundred Archon-class carriers and 170 Supercarriers formed the thin surface of the bubble. *Inside* this

bubble of slowly orbiting carriers is where he would position his Titan and supercapital forces.

The reason for this has to do with a quirk of how ships collide in space in *EVE Online*. Titans are extremely large ships, and CCP never thought there would be so many of them occupying the same space at the same time. When a group of Titans warps in on a single location all at once, they take up so much space that many times a few Titans will smash into each other and the janky video game collision will cause them to be bumped out of the formation and shot away into space. Slingshotted far away from the rest of the fleet these ships are easy prey. This was one of the primary ways that Titans wound up getting killed in fleet engagements: becoming separated from the healing carriers.

The 670 carriers and supercarriers literally surrounded the Titan fleet with a bubble of friendly ships, so if the Titan bounced out of the center of the fleet, it would collide with a carrier and the carrier would be shot off into space instead—a far less expensive loss.

The N3/PL Titans could fire their Doomsdays from the safety of the center of the Wrecking Ball sphere, secure in the knowledge that if the CFC/RUS fleet focused fire on one of them, the carriers on the perimeter would offer swift repairs. At the head of the force was Sala Cameron's Ragnarok-class Titan fitted with expensive components to draw the enemy fleet commander's fire, but also fully decked out with maximum armor and resistances to withstand as many Doomsday blasts as possible. Each of the 70 Wrecking Ball Titans was paired with a partner, and when the battle commenced they would work in teams to one-shot CFC/RUS Dreadnoughts. It was the most expensive race in *EVE Online* history.

The Wrecking Ball was the most expensive fleet ever built in *EVE* at the time, and only an ultra-wealthy group of older players like N3/PL could actually afford to field one, let alone pull off the logistics necessary to make it actually work. Anyone who has done a five-person dungeon in *World of Warcraft* knows how hard it is to get just 5 people to cooperate effectively. The Wrecking Ball fleet was an operation of incredible complexity that relied on the cooperation of literally hundreds of individuals to precisely position their ships and follow orders.

This was the marvel of fleet engineering that CFC/RUS hoped to kill. The CFC hated Slowcat carriers. It grew to loathe the Slowcat during the Fountain War, and so for the Halloween War The Mittani prepared an unstoppable force to break the immovable object that was the Wrecking Ball. In reference to one of the CFC's first fleet doctrines "Alphafleet," the new fleet style would be called "Omegafleet."

"We are going to smash our enemy's precious golden toys and rub their faces in shit," The Mittani wrote in an alliance update on December 15, 2013. "Here's how."

The CFC's "Omegafleet" was much simpler than the Wrecking Ball. It was a huge fist of Dreadnoughts, as many as could be found.

The CFC's Lazarus Telraven rose to meet the Wrecking Ball in HED-GP. He would put it to the test with an unprecedented 700 Naglfar-class dreadnoughts, capable of nearly one-shotting a Titan with their combined power every weapon cycle. Their mission here was to clear the Slowcats off the field, and pop the bubble of carriers that protected the Titans. It took the firepower of about 70 Dreadnoughts to one-shot a Slowcat before repairs kicked in from the other Slowcats. The open question was whether 700 Dreadnoughts could clear 70 Titans supported by 670 carriers, before those Titans cleared a critical mass of the Dreadnoughts.

Three hours prior to the battle, Manfred Sideous ordered his fleet into HED-GP, and the Wrecking Ball began to take shape. A cynosural field was lit, grew into its distinctive ball lightning, and from it emerged the 170 Supercarriers and 500 Slowcat carriers. The carriers began their slow process of moving into orbit around the cynosural field, eventually evening out to form a thin spherical layer a few dozen kilometers in diameter.

In flashes of bright pink light, dozens of Titans emerged through the cynosural field in staggered numbers. In the chaos, a few of the Titans collided with each other and bounced away from the center of the fleet, but collided with the bubble wall of carriers and bounced safely back toward the center of the formation.

The Wrecking Ball was working and the fleet was ready. Now Manfred Sidious would wait for Lazarus Telraven to make his move.

NODE STABLE

"Surprisingly, RUS/CFC were chomping at the bit for this fight," wrote PL's Elise Randolph, describing how the enemy fleet was specifically equipped to dismantle a Wrecking Ball. "All 700 Dreads were Doomsday tanked with a close-range fit – the most downright punishing Dread fleet assembled in recent

Above: One of the fleet's Titans is bounced away from the cluster and separated from the pack.
Opposite: A portrait of the avatar of Manfred Sideous of Pandemic Legion.

Above: The avatar of Vily, one of the lead fleet commanders of the CFC.

memory. This was augmented further with ~30 Titans and ~150 Supercarriers. Capable, for sure, of putting the PL/N3 Wrecking Ball fleet to the fire."

"These types of Capital slugfests are the unicorns of *EVE* PvP combat – a Fleet Commander's dream. Killing an N3/PL Wrecking Ball fleet would show once and for all that the Kings of Supercapital Combat were in fact mortal. On top of that, a punishing loss would simply cripple N3/PL for a non-insignificant amount of time before the ships could be replaced."

During previous wars there was rarely a single battle which could drastically hurt the finances of a coalition. Even in the Northern Coalition's war with the Drone Region Federation, when dozens of Titans were destroyed across several battles, the losses were still usually recouped. But now these coalitions were truly beginning to play with fleet concepts that were so valuable and powerful that they couldn't just be replaced. A winning strategy in this battle could very well determine the course of the game's history. Lazarus Telraven had a simple plan for how to defeat

the Wrecking Ball. He was going to warp all 700 of his Dreadnoughts directly into it.

"The RUS/CFC fleet had the numbers," wrote Elise Randolph. "Their plan was simple – come into system with subcaps and cyno their close-range Dreadnoughts into the PL/N3 Wrecking Ball and then unleash havoc in what would be the largest and bloodiest capital ship fight *EVE* had ever seen. A crude plan, perhaps, but no doubt effective. All that was left to do was execute."

The only issue was that a big part of the CFC/RUS subcapital fleet was already in the system, and so was the N3/PL Wrecking Ball fleet. CFC/RUS were not ignorant of *EVE*'s history, and knew that it was a risky game to jump 700 capital ships into a system that already has nearly 3000 people in it.

"The RUS/CFC fleets either forgot the lessons of old or were simply overcome by greed and took the daring approach to cyno all of their Dreads on grid," wrote Elise Randolph. "Not only on grid, but actually inside the PL/N3 bubbles. Brazen. The call was made, the cynos deployed, and the Dreads began to jump.

"On the brink of Supercapital Armageddon, all hell broke loose."

WRECKING BALL

At 16:45 GMT on January 18, 2014 Lazarus Telraven emerged from a cynosural field at the head of 700 Dreadnoughts specially tanked to survive Doomsday blasts and equipped with close range guns to take down the Wrecking Ball. On all sides the fleets were surrounded by the familiar blue hue of warp disruption fields, which wobbled like colossal soap bubbles and coursed with thin veins of electricity. Seventy huge Titans—each kilometers long—loomed nearby, while beyond circled the mass of carriers that formed the capital ship shell of the Wrecking Ball. The fleets were committed now, and there was no escape. Pandemic Legion fleet commander and frequent scribe of battles Elise Randolph wrote in the aftermath of the battle:

"PL/N3 command channel lit up; 'what the fuck did they just do?! They're cynoing into our bubbles with their subcap fleet on grid?!' The shock and awe would have to take a backseat, though, and PL/N3 quickly went to work breaking the fleet down into 5 distinct cores each shooting their own target. If they were to win they absolutely had to clear Dreads at a frenetic pace. This would be the only play PL/N3 had: clearing DPS before losing all of their [repairing] power. The PL/N3 pings went wild: 'Fight is on, when you die [self-destruct your pod to resurrect at the resupply at] Amamake and come back in a new Dread - cyno chains up.' Fast reshipping was the best hope to clear damage, and the cache of Dreads and carriers was ready to be depleted. Organized chaos from N3/PL saw reports that [we] were cutting through the RUS/CFC Dreads with ease - the [Doomsday] tanks weren't effective and Titans—which were split into pairs—were able to single-shot Dreads.

Hearkening back to the old days, strange things began to happen. The RUS/CFC Dreads would simply vanish after taking damage. Others would remain invulnerable. The [RUS Dreadnoughts were] visually stuck in a warp-tunnel from [the RUS pilot's] perspective, and were materializing in HED-GP and then being magically transported back to where they came in F4R2-Q. The node creaked to a halt, 25 minute module lag from the PL/N3 perspective [editor's note: "module lag" is the amount of time it takes in extreme lag for a player to give their ship a command and for the server to execute that command in the game world.] Drone assign simply broke. Credit to CCP on a day where most will blame them, the

node did stay functional in the face of 4000 people on grid. The RUS/CFC Dreads were showing up in clumps of 100, vanishing again before dying. After about an hour of this confusion (six minutes of elapsed [in-game] time), all of the RUS/CFC Dreads began to materialize in earnest. The 25 minute module lag however, meant that the relatively uncoordinated Dread fire was relegated to easily tankable splitfire.

The only thing doing significant damage was the Titan Doomsdays - after all 25 minute module lag is irrelevant with hour-long cooldowns. Jumping onto a grid with 2700 people, of which 1800 are yours, is a rookie move by all accounts. It is what saw Pandemic Legion lose the then-largest capital fight in Y-2ANO some five years ago - a fight that the CFC were involved in. The same type of fight that RUS experienced fighting the Northern Coalition in Uemon and O20-2X some three years ago.

So a completely abysmal execution changed what was a decent chance at killing a Wrecking Ball fleet into an utter turkey shoot. RUS, dejected by this catastrophic failure, completely gave up on calling [targets.] The CFC, equally frustrated at the failed execution, at least tried to make the best of the situation and killed [warp disruption bubbles] allowing them to extract the bulk of their Dreads. Meanwhile the Black Legion component, the ~surprise~ 200 dreads, simply opted to go home instead of showing up to the fight.

Some three-hundred and fifty RUS/CFC dreads died between HED-GP and various other systems they panic-jumped to. In

Above: The CFC's 'Omegafleet' is repulsed at HED-GP as lag turned the battle into a nightmare for the coalition dreadnought fleet. Opposite: The N3/PL Wrecking Ball seemed indomitable following its impressive performance in HED-GP.

return for this hefty price, a mere ~10 N3/PL [capital ship losses.] [...] The old school strategy devised by Manny, Vince [Draken], and the rest of the N3/PL Fleet Commanders paid dividends albeit in a very anti-climatic and drawn-out way. At the end of the day the largest Wrecking Ball fleet ever assembled was able to execute perfectly and completely and utterly devastate the CFC/RUS Dread fleet, proving once again that N3 and PL use capitals better than anyone else.

In the current *EVE* climate a 350 Dread loss is not coalition-breaking by any means - the fighting will continue and the south will be ablaze with action. [...] The fighting is far from over, though it will be months before we see a capital fight of this scale again.

— Elise Randolph, Pandemic Legion
January 18, 2014

The shaky CFC/RUS coalition quaked as the CFC cast blame and vilified Black Legion for staying out of the fight, saying it was more evidence that Elo Knight was a treacherous mercenary.

Meanwhile, confidence swelled among the N3/PL fleet commanders, who saw this as vindication at long last that the ClusterFuck Coalition was *EVE's* aging, decrepit empire and they were its true superpower.

However, the reality check would come in just seven days.

B-R5RB

Throughout the Halloween War, N3/PL had been staging out of a station in the star system B-R5RB in the region of Immensea. Nulli Secunda were the owners of this critical offensive base, but the campaign commander of the combined N3/PL forces at the time was Pandemic Legion fleet commander, Manfred Sideous. This meant that Manfred Sideous had to get permission from someone in N3's Nulli Secunda whenever he needed to use the system's infrastructure. In particular, there's a critical piece of infrastructure in nullsec called a "cyno jammer" which makes it impossible to deploy cynosural fields in a system. Since cynosural fields are the only way to get capital ships into a system, this is an essential tool for defending space. The cyno jammer needed to be active during Russian prime time to keep the system safe from an onslaught, but Manfred Sideous also needed to be able to deactivate the jammer so he could warp his own fleets in and through B-R5RB to successfully coordinate the defense. Since the system was owned by Nulli Secunda, Manfred Sideous had to ask a Nulli director (if any were online) to deactivate the cyno jammer and then turn it back on again once he'd left. It was a procedural annoyance, and a common one in logistics-heavy New Eden.

The other issue was that Pandemic Legion didn't fully trust Nulli Secunda to keep this crucial system secure after the embarrassing disbanding of its rental program during the Fountain War.

The problem was that if Nulli Secunda simply gave control of the star system to Pandemic Legion, all of the defenses would've gone offline as soon as Nulli Secunda dropped and transferred their claim. It would then take Pandemic Legion four weeks to get back up to full sovereignty defense bonuses, which would have left them incredibly vulnerable in the meantime.

The solution was for Manfred Sideous to work a little paperwork magic. Like a true nullsec alliance leader he was intimately familiar with the ins-and-outs of the *EVE* user interface. He knew all the obscure functions, all the trickiest bugs, and, most importantly, he also knew all the administrative ways that shell corporations could solve difficult problems.

The problem was easily solved by Manfred Sideous setting up a dummy corporation called HAVOC with only one member: himself. Then Nulli Secunda invited HAVOC to be a member of its alliance, and assigned it to have control of the structures in B-R5RB. Voilà. Manfred Sideous could now control the cyno jammer with an alternate dummy character without having to transfer the sovereignty claim. A solution as elegant as the problem was boring.

This was a common logistical workaround that every alliance had done a dozen times by this point. It wasn't really a special or unique arrangement, and yet it is arguably the single critical choice that shaped the game for the next two years.

What happened next is disputed. Manfred Sideous says he packed the HAVOC corporation wallet with enough ISK to cover four months of sovereignty bill payments and clicked autopay so he'd never have to think about it again.

The Tranquility server disagreed. According to the server, Manfred Sideous did not click autopay. On January 27, 2014 the sovereignty bill for control of N3/PL's forward staging base went past due. ●

THE BLOODBATH

"Here lie the wrecks of monstrous ships, commemorating a battle
that blotted out the sky on Jan 27-28 [2014]. Two coalitions of
capsuleers clashed in vessels numbering in the thousands, causing
destruction on a scale of war never before seen by human eyes."
— Monument erected in B-R5RB

A t 11:00am in London on January 27, 2014, the *EVE Online* Tranquility cluster entered its scheduled daily downtime. As it did every day, it ran a series of checks to determine whether the state of the game had to be changed before the next cycle began.

There are thousands of star systems with hundreds of starbases and outposts, and the Tranquility server has to keep track of all of it. It also has to serve as the banker that collects regular fees from the players for thousands of types of daily interactions.

One of those split-second jobs is to determine the rightful owner of every nullsec star system. If a player group wants to control a star system they also have to pay a small price in ISK that's collected by the server. It's not a heavy cost to a major nullsec alliance, but it's a token acknowledgement that in the lore of the game world the player alliances are paying tithe to the in-game NPC factions. (In reality it's what game designers call a "gold sink," flushing a regular supply of currency out of the game to keep inflation under control.)

The Tranquility server has to collect those payments and make sure everyone has paid their bill for every star system. Since the cost is not terribly burdensome, most everyone just clicks the little box that says "autopay" on the sovereignty management interface. The server collects the money every month, and the alliance gets to maintain sovereignty and continue to conduct their business.

It was a completely normal day in a star cluster that seemed like it might grow stagnant following the disaster the CFC experienced in HED-GP. As downtime

Opposite: A small view of the battle which would become known as the Bloodbath at B-R5RB.

continued, nobody would've guessed that today—just a week after the most legendary defeat in years, and the one year anniversary of the impromptu Battle of Asakai—was the day of *EVE's* reckoning.

Today was the day of the battle which would define *EVE* in the minds of the public for a half decade (and counting); the only digital event which has truly eclipsed the disbanding of Band of Brothers in its shaping of the *EVE* global reputation.

It may prove to be the most climactic and important single battle that ever occurs in *EVE Online*, past or future. If so, that would also make it the most consequential battle in the history of online gaming so far. The battle lasted nearly 21 hours, and involved more than 7000 players from hundreds of corporations all across *EVE*. Ever since, players have referred to the events of January 27, 2014 as The Bloodbath at B-R5RB.

DOWNTIME

After thirty minutes down for its scheduled daily maintenance, the Tranquility server came back online at 11:30am and began accepting connections.

It was an unusually quiet cycle by nullsec standards. On any given day there are often as many as twenty or thirty changes to the enormous, star cluster-sized game board that is New Eden. Today there were only two. A small alliance called The Unthinkables gained a star system in the region of Scalding Pass, and Nulli Secunda dropped its sovereignty claim in B-R5RB, the forward staging base for both N3 and Pandemic Legion.

Manfred Sideous claims he is certain he checked and double-checked the autopay box, and that the event is eerily reminiscent of a bug he'd previously encountered in the sovereignty payment system a year prior. To be fair, *EVE's* user interface is notoriously buggy.

Though this was a disastrous turn of events, there were no red flashing lights or alarm signals. It simply happened. Server downtime happens at 11:00am GMT for a reason, it's the time when the population is at its daily low, pre-dawn hours in North America, morning in Europe, but Russian prime after school/work hours. Thus, it was a few hours before the community really grasped what had happened.

Nobody at this time knew anything about what was about to occur. Manfred Sideous only noticed the dropped sovereignty claim when he logged in shortly after downtime to defend a Drone Region outpost that was scheduled to come out of reinforced mode. He immediately dispatched a new Territorial Control Unit to be anchored in B-R5RB to renew the sovereignty claim. However, TCUs take eight hours to come online. With the TCU anchoring, Sideous waited to find out whether this bizarre mishap might go unnoticed by his enemies.

It, of course, did not. Within the hour, a CFC scout was sent out to figure out what was going on in B-R5RB, and returned with word of an onlining Territorial Claim Unit being guarded by a small defense force.

THE BOWELS OF TELRAVEN

The highest-ranking CFC officer online at the time was Lazarus Telraven. Telraven was an American Fleet Commander, and he had only logged on because he couldn't sleep due to a bad case of stomach flu. Telraven was conferring with the other leadership that were online, including Sort Dragon (whose geoposition in the Australian timezone made him an ideal link between the English and Russian factions of the RUS coalition) trying to understand how to interpret this turn of events. The Russians thought it was an elaborate trap, but Sort Dragon wanted to go in as quick as possible and exploit this unlikely gift. Nobody was yet aware that the sovereignty had dropped because of a simple clerical error—or a bug, depending on who you believe.

"I woke up that night pretty sick," Lazarus Telraven told Crossing Zebras interviewer Xander Phoena. "I think I went to bed at like 1, I woke up at like 3:30, 4 o' clock and that's when I found out about it. I actually

took the day off work for illness, but I ended up being stuck on *EVE Online* for like 20 hours or whatever."

He was trying to wake up The Mittani with a call to his cellphone, and fleet commanders Vily and Mister Vee were at work and couldn't get home. While he waited for permission for a full supercapital escalation he sent in a contingent of siege dreadnoughts and a subcapital support fleet to make the CFC's presence known. Sort Dragon, whose alliance was now working with the Russians, was already there.

"B-R happened because of me," Sort Dragon told me in the defiant, heavily slanted way he often did when defending his legacy in *EVE*. "I had the Russians. I was the Russian leader at the time. Mactep and [another Russian leader named Union] said, 'Sort speaks for us.' I was given the option to engage. Manny fucked up the TCU, everyone knows that. But the decision to enter the field was mine. Because Laz didn't have the balls to do it. And Laz, I basically said to him, 'I'm going in, let's go for it.' And Laz is like, 'Oooh I'm not sure, I need to talk to The Mittani, Oooh I'm not sure." And I said, 'Well I'm already in, you need to get here."

A WIRED.com article written in the days afterward indicated Lazarus Telraven arrived in B-R5RB at about 13:00 GMT at the head of 300 sub-capitals and 45 Dreadnoughts.

With a small but dominant force in the system, Telraven and Sort Dragon dispersed the defense force, destroyed Nulli Secunda's Territorial Claim Unit, and began anchoring another. To make matters worse, with no official sovereign owner of the system, the station within that system could be conquered by anybody who shot it. While the TCU was being destroyed, a contingent from RAZOR Alliance claimed the station in B-R for the CFC, preventing any N3/PL ships from docking or leaving the station.

Manfred Sideous began to panic, and rushed back home from the Drone Regions to take personal command over the home defense. He later said that he wanted to quickly rally the biggest possible Titan force he could, to try to scare the CFC back out of B-R5RB. Sideous quietly rallied everyone he could find who owned a capital or supercapital ship and told them to make their way to the staging position in B-R5RB, and begin assembling the Wrecking Ball.

Lazarus Telraven finally roused The Mittani, who advocated caution.

"I called his phone and woke him up when things looked like it was going to escalate because when you need someone to bang drums to get [lots of pilots to a battle] you get mittani," Telraven wrote in a Reddit AMA (Ask Me Anything) the next day. "He was [skittish] and hesitant and I think came close to trying to get me to not go in but had he told me not [to] jump I would have jumped anyways."

With Mister Vee still not responding to Telraven's text message, Laz wrote one last text,

"I'm going all-in. Get here."

ALL-IN

The Mittani began the mustering of the Goons. Telraven got everyone into line, and organized three fleets to take down the most feared fleet commander in 2014 New Eden: Manfred Sideous, the architect of N3/PL's "superweapon."

For years, the two great machines called N3/PL and the ClusterFuck Coalition had been building their ability to summon, coordinate, and transport thousands of individuals to a single location in a virtual world under a coherent command structure. Their industrial teams had been building ships with the expressed purpose of one day winning a huge Ragnarok-style nullsec confrontation with the opposing bloc. And it all came to a head here at B-R5RB, a system which was tranquil just a couple hours ago and yet would soon become synonymous with the mother of all battles.

Time Dilation kicked in as the hundreds of players in B-R5RB surged into thousands. Both sides called up every ally and searched their contacts for anyone else with a Titan who could help push this battle in their favor. Lazarus Telraven ordered the CFC's Titan fleet into the battle, and from then on they were committed.

For Manfred Sideous, a sickening realization set in quickly when he saw the first CFC Titans appear on-grid.

"I didn't think they would escalate with supercarriers and Titans and everything was going to be OK for me like it had been 100 times before," Sideous said. "But it wasn't."

"As soon as they dropped Titans, I knew it was going to be bad from there on," Manfred Sideous said in his post-battle interview. "Because the thing about [Time Dilation] is while the rest of the server is going along plugging away at normal time, in that system only seconds or minutes of [real] time [pass.] So pretty much it allows everybody to keep throwing things into the fight if they so choose. I knew that since they had committed Titans already, that they would be willing to commit everything that they could possibly bring to bear. At that point I was simply trying to do the best with the situation that we could and take it from there."

Both fleets were stuck in a wide cloud of anti-warp bubbles, and the two sides would try to hold their ground until some sort of tipping point was reached.

"Unlike nearly every other large scale supercapital engagement up till this point, both sides thought they could win," reads a development update published by CCP Games after the battle. "They continued trying to get every single pilot into system with the most powerful warships they could bring to bear. After a few hours, the field was being lit up by Doomsdays and the glittering hulls of hundreds of Titans and supercarriers and thousands of Dreadnaughts and Carriers and smaller ships."

Below from left to right: The avatars of Manfred Sideous, Sort Dragon, and Lazarus Telraven.

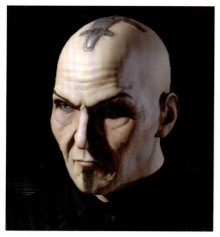

Above: A player's perspective as the battle was underway. Below, and Opposite: Actual screenshots from the Battle of B-R5RB on January 27, 2014 taken by popular *EVE* figure and filmmaker Lord Maldoror.

The main issue at this point was the time. Many of the biggest *EVE Online* battles prior to this only stopped because the server went down for daily maintenance, and whichever side was losing usually opted not to log back in after the server came back up. Because the events in B-R5RB had occurred right after downtime, there was still some 20 hours to go before the next downtime arrived. It was an unprecedented amount of time for these two coalitions to escalate the fight.

The sky in one direction was stained by the red starlight of B-R5RB, and in the other by the pale blue-green skybox of the Immensea nebula. Between them, a densely-clustered foam of ships and warp bubbles to fill the field of view as approximately 7,548 *EVE* characters became clustered and tangled in a singular melee to define the future of the game. The two fleets clashed together in a brilliant cloud of ships, drones, missiles, repair lasers, and every other particle effect Tranquility knew how to generate. If ever there was a moment that could capture *EVE* in a single screenshot, this was it.

Though the battle had been spontaneous and escalated wildly within hours the stakes were now utterly dire. The opportunity for the CFC/RUS Coalition was to capture the headquarters of the only real nullsec resistance to its dominance. And for N3/PL this was now a battle to avoid complete oblivion, and yet, a slim chance remained. The CFC's super fleet was out in the open. If the Wrecking Ball held strong, the tables could be turned entirely.

This whole game—every haul of ore, laser blast, or military invasion—had been building for eleven years to this one moment on January 27, 2014. This moment, nearly frozen in time by Time Dilation and lag, is the perfect distillation of the story of *EVE*. The screenshots of this moment—the peak of the great 11-year contest—look less like images of a video game, and more like mournful renaissance paintings, if only the masters had painted starships.

After four hours, CFC/RUS had claimed eight of N3/PL's Titans while losing seven of its own. The cost already topped two trillion ISK, nearly twice as much as any battle in *EVE Online* history. Fifteen Titans down. Sixteen hours to downtime.

DOOMSDAY CLOCK

After four hours of this slugfest the two sides stood roughly even, but Manfred Sidious knew that things were about to get much worse. After hours and hours of exchanging Doomsday blasts, American prime time

Above: A full panoramic view of the peak of the greatest battle at this point in the history of *EVE Online*.

was coming up, which would mean an influx of pilots for both sides, but moreso for the CFC.

"It becomes a game of managing Doomsdays," Lazarus Telraven wrote in his AMA after the battle. "Each individual fleet commander is relaying to me who has available Doomsdays, and I coordinate who should shoot at what target."

As the CFC's American pilots logged in, Lazarus Telraven had to figure out how to use the hundreds of fresh pilots to tighten his grip on the Wrecking Ball. He noted that the system was already full of more than 4000 pilots and could potentially be reaching a breaking point soon. He didn't want to bring so many people here that their numbers crashed the server and allowed N3/PL to escape. Instead he dispatched most of his subcapital forces to surrounding systems to intercept N3/PL pilots when they re-shipped and made their way back to the battle. By doing this, Laz was able to spread some of the server load out to surrounding systems, while also accomplishing a crucial strategic task.

With every hour that passed, the CFC's advantage grew, and N3/PL could see the tipping point fast approaching. With the influx of pilots now coming online for CFC/RUS there would soon be no way to turn the tide. Manfred Sideous decided that the only way to turn things around would be to disrupt communications between CFC and RUS and hopefully split their damage. That meant blasting Sort Dragon's Avatar

flagship off the field. Taking him out of the fight was a low-percentage hail mary, but it was the only plan Manfred Sideous could see that might plunge CFC/RUS into disorganization.

More than half a dozen Titans from within the Wrecking Ball began to glow and shimmer in slow motion before sending lances of laserfire across the battlefield at a light speed crawl, converging in slow motion on the hull of Sort Dragon's Avatar.

However, Sort Dragon had the foresight to overheat his ship's shield hardeners about 20 minutes ago in the blinding lag in preparation for theoretical incoming damage—meaning he could tank far more damage than usual. He pushed the button twenty minutes before anybody fired at him, but in full Time Dilation that was bizarrely perfect reflexes. When the blinding beams fizzled, Sort Dragon's Avatar was badly damaged but still standing. More damage rained down upon Sort Dragon to try to tip his hull past the point of no return, but by now the CFC could see he was being primaried and diverted their triage ships to prop him up.

While the full artillery of the Wrecking Ball rained down upon Sort Dragon, Lazarus Telraven continued organizing CFC Doomsday volleys, making sure 3-5 Doomsdays hit each N3/PL Titan to ensure it would be killed in one volley. By the time Manfred Sideous managed to finally down Sort Dragon's Titan, Lazarus Telraven had taken down five Pandemic Legion Titans in return.

The number of Doomsday weapons was the all-important factor, and 11 hours into the battle N3/PL had lost 18 of them and taken down just 8 in return. A battle of attrition was being waged. Not only was N3/PL's fleet dwindling, but the CFC's was still getting stronger as more Americans continued to log on.

RETREAT

Around 23:00 GMT—twelve hours to downtime—N3/PL lost the critical mass of Doomsdays to be able to reliably one-shot CFC Titans at a reasonable pace, and hope for the battle slipped beneath the horizon. Manfred Sidious handed off command after a long night on duty. The situation in B-R5RB turned from a battle into an evacuation. But the scale at which Sideous had committed to this battle meant that was a tremendous undertaking that would take the rest of the night.

The Wrecking Ball was broken, and there were still twelve hours left until downtime. Even in the slow motion of Time Dilation, that was an eternity in which to absorb volley after volley from what had now become unquestionably the most powerful supercapital fleet in New Eden. Just a week ago N3/PL saw the battle at HED-GP as a sign that its star was rising. Now it was hoping merely to survive. The cost of the battle had now surged past 4.5 trillion ISK, worth more than $100,000 USD in January 2014. The loss was already nearly four times greater than any battle that had ever occurred in *EVE* before.

"N3/PL fleet commanders sounded the retreat. The objective was now to save the expensive and difficult to replace supercapitals, sacrificing the regular capital ships if needed. [...] They switched their fire to dreadnoughts, determined to take anything they could with them in death. The last Titan loss on the RUS/CFC side would be Chango Atacama from the CFC alliance Circle-Of-Two."

N3/PL began trying to extract their ships as methodically as possible from B-R5RB, but once they started doing so the CFC was free to open fire without restraint. CFC sky command switched many pilots into Heavy Interdictor ships. They wanted as many N3/PL ships stuck here as possible until the clock reached 11:00 GMT again and the server went down.

Throughout the night the battle spiralled into a chaotic mess. N3/PL attempted to keep calm and methodically extract as many ships as possible from the system now that the battle was lost, but the extraction was slow, and the night was long.

The server struggled to calculate and display the thousands of wrecks which flickered in and out of existence in the crushing lag. The space near the outpost was clouded with anti-warp bubbles and crisscrossed by thick red and green Doomsday beams—as the remnants of the N3/PL fleet attempted to organize the extraction.

Slowly, brutally, *EVE Online* stuttered and lagged its way through the night toward the Bloodbath's grim final resolution. ●

IMPERIUM

"And to downtime the battle went. The CFC/RUS capital fleet continued to kill capital ships right up until the black screen. Predictably, N3/PL gave their members orders not to log back in after the fight and the most expensive battle that *EVE* has ever seen abruptly ended."

— Alizabeth, TheMittani.com

During the final twelve hours of battle, N3/PL had managed to kill 5 additional CFC/RUS Titans, nearly as many as any side had ever killed in a previous confrontation. But it came at devastating cost: an unheard of 59 of its own. Having lost a Titan in B-R is today a point of great pride for some pilots.

Taken together, one quarter of all Titans lost in *EVE Online* history were destroyed in B-R5RB on January 27, 2014, and the vast majority were destroyed by the combined CFC/RUS fleet. Even as the victors, CFC/RUS absorbed more damage than had ever been taken by both sides in any battle. The cost of its losses alone topped 2 trillion ISK. But you should see the other guy. Total losses for N3/PL: 8.5 trillion ISK. The brunt of the losses were absorbed by Pandemic Legion: 4.5 trillion. NCdot was next at over 3 trillion. The historic destruction at B-R5RB eclipsed that of the massacre at Uemon by 8 times.

Nothing close to this had ever happened before, and both winners and losers were caught in the afterglow of a gaming event that captured the world's attention. Both side's fleet commanders hosted Reddit AMAs and made media appearances. But Sort Dragon largely didn't, because he became...busy IRL.

"Two weeks after B-R I went to prison," Sort Dragon told me. "And I came out three months later to find out that Laz was 'The Butcher of B-R5RB.' TMC [TheMittani.com] controlled the narrative at the time, and they told everybody that he was the Butcher of B-R and everything, except he was throwing up in the toilet most of the day because he had food poisoning."

EVE reporters went into action trying to get the story straight about what had happened here. And as he so often did in this era, Elise Randolph of Pandemic Legion commemorated the event with an in-depth battle report.

"For the last few years we've been using the money we get from renting/[technetium]/contracts into building a capital armada of Dreads and carriers for a huge capital slugfest with the intentions of losing a full capital fleet and replacing it as the fight goes down. Just like with subcaps, the ability to reship in a capital fight is hugely important. The B-R fight, though, wasn't meant to be the site of a huge capital brawl. A week earlier in HED-GP we were prepared to lose everything, reship, and go back at it. Luckily those cynos and caches were still prepped and in place for the B-R fight so we were able to reship for the biggest capital/supercapital fight

THE IMPERIUM

Above: A wider view of the Titanomachy monument with the statue situated at center and surrounded by wrecked Titan hulls.

EVE has ever seen. From that end, the capital loss we incurred has been totally replaced as fast as people can accept [construction] contracts.

What we did lose, however, was a huge amount of active Titans. Titans have very few uses from a combat perspective except when it comes to capital fights. In these cap fights Titans are king and essential; as we saw in HED-GP they are very daunting and tend to win fights by intimidation. Jumping into 50 Titans means you risk losing huge assets every Doomsday cycle.

The cost to replace these suckers in the amount that we lost is certainly not trivial—a few trillion ISK—though certainly replaceable. In the short term it will limit our ability to create these capital fights (which I'm sure everyone is thankful for) and it changes how we approach fights.

Perhaps five years ago a loss of this scale would send an alliance into a tailspin, though the cold truth is that the top tier of alliances in *EVE* these days can incur a multi-trillion ISK loss and keep on chugging. A week ago CFC/RUS lost 350 dreads and fielded the same exact fleet on Monday [at B-R5RB.] The ISK lost from the B-R fight is a much greater scale, of course, but the ISK is still there for everyone. The lost momentum in the Southern conflict due to the B-R fight is a far greater loss than the ISK assets. Because try as you might, you cannot buy swagger."

— **Elise Randolph,** Pandemic Legion
January 28, 2014

Pandemic Legion Fleet Commander Grath Telkin posted a reply of his own to show what was next for Pandemic Legion, explaining that he didn't blame his allies for the loss, but that he needed to pull back and get his alliance back on its feet.

"I don't throw allies under a bus, Vince [Draken, leader of NCdot] is some-one I consider an e-friend, and despite how things go with him and other people we have dealings with, Vince has always been there and always had our back.

That said, my corp, my fucking corp, took a hit of around 1.5 trillion ISK. My alliance as a whole took a shot to the nuts somewhere in the neighborhood of 4.5 trillion ISK. Those numbers are beyond staggering in nature and dwarf

the amount of money more than half the alliances in *EVE* will ever see, much less recover from. And right now that's what needs to happen. I need my people to get a chance to recover.

I've been in PL for a really long time, and all the welps we've had have been chump change to the actual members in comparison to the losses we're facing here, and coupled with Black Legion knocking on the door in [the Drone Regions] near constantly I made a judgement call to get my people whole.

I eat a shit sandwich, I pull back, I get my people right, I get my house right, and I get my cap fleet back in order, and then we see where life is and how things are going.

I don't expect people to like my call, and personally your opinions on my call mean less than fuck all, what matters to me is how my people are doing and dealing with things and how the future of the alliance looks, and the best way for me to do that is to back our shit up to our house and have some Legion on Legion combat up north.

Anything outside of how the alliance recovers is irrelevant to me."

— Grath Telkin, Pandemic Legion January 28, 2014

The morning after the brawl at B-R5RB reports began to spread worldwide about the incredible spectacle that had unfolded within *EVE Online*. The great clash of plasma and tritanium was boiled down to one fact: if you were to convert the total sum of ISK lost in the Battle of B-R5RB to United States Dollars at the going exchange rate then the battle bore a staggering cost of more than $300,000.

It was the most destructive day in the history of virtual spaces. Finding information about the year following B-R5RB in *EVE Online* is far more difficult than other eras simply because the search results are coated in a thick layer of news pieces about B-R5RB. The media fawned over the event for months as CCP Games wisely co-opted it for its marketing endeavors.

The narrative that emerged in the press cast B-R5RB as the latest battle in an ongoing war with unknown consequences for the future of the conflict. But within *EVE Online* the story was different. Inside *EVE*, the impact of the battle was felt immediately as the N3/PL coalition recoiled from an apocalyptic welp. The worst welp, in fact, in the history of welps. The star cluster now had to grapple with two facts: 1) The Wrecking Ball was not a suitable counter to being massively outnumbered by supercapitals, and 2) the CFC super fleet which was already far larger than N3/PL's before the battle was now even larger and the advantage was growing.

On top of the massive losses of ships, N3/PL had also lost B-R5RB itself and with it its forward staging base. After the battle the coalition was falling back, and the CFC was in hot pursuit. N3/PL retreated to a new station but were swiftly caught. The imposing CFC supercapital fleet—now unquestioned in its dominance of the star cluster—moved in and set up a great camp that kept N3/PL trapped within its station.

For a week following the disaster of the battle, the ClusterFuck Coalition maintained an overwhelming presence over the station, bottling up the only other legitimate opposition in New Eden.

But then, abruptly, The Mittani declared that the Halloween War had been won, and his fleets would be returning to Deklein in victory. The situation spiralled as both of the warmachines that emerged to fight B-R5RB started to cool. Not only was N3/PL reeling, but the CFC's arrangement with Black Legion had now deteriorated beyond repair, and the RUS coalition was going through myriad internal difficulties as well, not in least part because Sort Dragon was now serving a prison sentence. Even the core of the CFC was showing increasing strain.

With the coalition frayed there was no way to officially extinguish N3/PL, and the badly fractured superpowers instead eased into a new cold war. The CFC could not officially wipe out N3/PL, but after a cataclysmic loss of Titans and the sundering of the Wrecking Ball, N3/PL could not even threaten the ClusterFuck Coalition in any meaningful way.

"As the month came to a close, with the consequences of the battle of B-R5RB beginning to make themselves known, an uneasy Cold War began to settle in the factions of the CFC and N3/PL," wrote a writer named Markonius Porkbutte. "Much of the galaxy set about finding room to manoeuvre around two very large power blocs as space for grabs was seeming more and more doubtful."

Immediately after B-R5RB, N3 and Pandemic Legion began a process of consolidating the sovereignty in the south in every territoriy beyond the CFC's reach. The two allies formed huge blocs of renter space as large as the empires of old but entirely for sale. Pandemic Legion's renter alliance Brothers of Tangra and Nulli Secunda's "Northern Associates." turned the entire south and west of EVE—historically the less desirable territories—into a renter farm.

The evolution of the EVE Online social system had all been leading up to this point. The most important factor for geopolitical dominance was the number of pilots that could be brought to a battle and the number of Titans a coalition could field. Now that there was no viable supercapital contender, faith in the game's ability to produce conflict began to erode. The two predominant power blocs had treaties promising not to invade the other. The BOTLRD Accords were still in effect. They couldn't outright destroy one another and there was huge amounts of ISK to be made in stalemate. So why fight? Fleets were still formed, and gudfites were still staged to keep the members entertained, but sources say the organic game of nullsec ground to a halt.

I've chosen to conclude the story of Volume II in this series here because it's clear that after the Battle of B-R5RB a new era had come about. The history of the next two years reflects that as the ClusterFuck Coalition became undisputed as the dominant superpower in nullsec.

There was a counter-invasion attempt by N3 in early 2015, but sources don't indicate it amounted to much. The game slowed down drastically, and the outer reaches of nullsec became largely static. For nearly a year afterward, the sovereignty maps from this era stay virtually identical, dominated by the huge renter empires of each coalition.

The problem was supercapitals. Their proliferation had tipped the balance of power, and huge networks of Titan jump bridges allowed them to leap around the star cluster practically at-will. Previously it could take hours for pilots to travel from one end of New Eden to the other. The Battle of B-R5RB was able to escalate so wildly because over the years the players had developed simple means of moving Titans to any corner of New Eden within about 15 minutes.

Many were relieved when CCP at last announced a new sovereignty system would be coming to EVE Online to shake things up, and the chief focus was new mechanics for supercapital force projection. In the "Phoebe" expansion, CCP introduced a "jump fatigue" mechanic that limited how far a single ship could travel in a given time period which drastically altered nullsec gameplay. But until that time, even small groups of friends often reported being surprised by Pandemic Legion's entire supercapital fleet, because there was simply nothing else to do.

"Too many killers will quite successfully chase away everyone else," Ultima Online designer Raph Koster wrote in 1998. "And after feeding on themselves for a little while, they will move on too. Leaving an empty world."

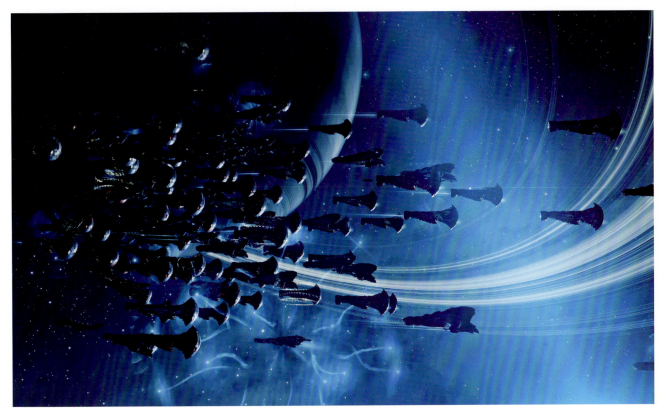

Above: In the era that follows the conclusion of this book supercapital proliferation continued unabated and produced many awe-inspiring battles that would've seemed impossible in 2010. Below: The logo of IWANTISK, an online casino-style website where players could gamble their currency.

THE IMPERIUM

In April 2015, after 15 months of complete dominance over nullsec, The Mittani made an announcement. The ClusterFuck Coalition was being restructured. Unfaithful allies were culled, and the coalition would be getting an official rebrand to something more befitting a galactic superpower of its stature. Henceforth, he announced, the organization would be known as "The Imperium."

"The CFC began as a rag-tag group of allies of convenience, thrown together by circumstance and geography in the aftermath of a power vacuum," The Mittani wrote in the announcement. "Who deserves the credit for transforming these disparate groups into the coalition which repeatedly won *EVE Online*? Vile Rat and his brainchild, Corps Diplomatique - and then after our friend's senseless murder, the tireless efforts of his protege, Sion Kumitomo. Under the guidance of Corps Diplomatique we have solved the problems of ego and byzantine politics which repeatedly have crippled our foes. [...] We have created a modern state in internet space- with highways, borders, a loose federated system of government, networked communication systems, and innumerable social programs. The ragtag clusterfuck of 2011 has grown into a true space empire - an Imperium."

Many in *EVE Online* were left to consider the journey that had unfolded on the Tranquility server over the past twelve years.

"The sad truth is that no one can contest the Imperium," reads a CrossingZebras.com editorial dated November 23, 2015 by author Tubrug. "The evolution of Goonswarm and the Imperium over the past decade is truly fascinating. GoonFleet originally established itself as being a carefree group of players fighting 'elite PVP' types and those who took the game too seriously. Truthfully, what is the difference between alliances in the Imperium in the present day and the very people they swore to destroy?"

It would not be until 2016 that the whispers of revolution were once again heard throughout New Eden. By that point, The Imperium's advantage was so drastic that many wondered who even had the cash to mount an attempt to break The Imperium's iron grasp on this game?

The answer, as usual, would emerge from *EVE's* seedy underworld: a grey market casino called IWANTISK.com and an amateur spy named LennyKravitz2. ●

EPILOGUE

"I am thus optimistic that historical studies of human societies can be pursued as scientifically as studies of dinosaurs—
and with profit to our own society today, by teaching us what
shaped the modern world, and what might shape our future."
— Jared Diamond, Guns, Germs, and Steel: The Fates of Human
Societies

The story of *EVE Online* is the story of hundreds of thousands of people's lives, lived digitally within one of the most complex and dynamic virtual social spaces ever built.

EVE represents the extreme of what you can experience in an online game. It's the edge of online exploration; a social space where many gamers are legitimately afraid to go because it has been largely unmoderated for most of its history. Step through *EVE's* login portal, and you'll find unfiltered humanity, in all its splendor and unexpected brilliance, but also its ugliness and sinister genius.

"It's like real life, but without rules," said Sort Dragon. "For the last 12 or 13 years I've been a leader in 0.0, and it's a very interesting insight into people's psyche, and how they are as people and what they believe in. You meet some very interesting people. When you take away the need to conform to rules they enjoy themselves a lot more, and you get to see into the deep dark soul of them."

By refusing to put limits on what players can do to seek advantage over other players, *EVE* allows its players to use their ingenuity to build human networks,

solve problems, and create systems. These are some of humanity's greatest tools, and *EVE* engages with them at scales no other video game ever has. However, those same tools can be turned for evil purposes.

The boundless room for human ingenuity results in a unique darkness balanced by a profound bonding among those trying to save a light from that darkness.

But the darkness remains. Complete freedom on the internet is a wish upon a monkey's paw, and it inevitably comes with devastating drawbacks. Humanity is all its own worst horrors. In any given virtual population, some people's goal will be to hurt other players. Their goal might even be to hurt a specific player.

EVOLUTION

Death and loss are at the center of *EVE's* gameplay design. This is true in some other games as well, but because of the way *EVE* is structured this also applies to player organizations.

Human social networks have always been the most successful tool in *EVE*, and that also means they are the number one threat to other player groups and the first thing to be targeted in a conflict. It didn't take long for *EVE's* players to begin looking into the human history of espionage to learn how to use spies and propaganda tactics that might help break a rival group's morale and social bond. Unfortunately, that impulse has also led to some players taking things too

Center: The modern day logo of *EVE Online*.
Opposite: A massive new type of structure planned for *EVE's* 2016 'Citadel' expansion, the Keepstar.

251

far and either threatening others in real life or revealing their personal information to the online masses.

The true game of *EVE* is a meta-game that exists outside the virtual realm, largely on chat servers and community platforms, and that is a wild, occasionally dangerous space that CCP has never figured out how to police.

"There are things that did cross the line," said Pro-GodLegend about the espionage underbelly of the game. "EVE is weird in that there's a grey area when it comes to spying and counter-intelligence. There's a lot of IP tracking, a lot of cookie tracking. There's some stuff that a lot of people would frown upon in just about any other situation *but EVE Online.* [...] I am slightly worried. I have a well known name in *EVE*, and there's a tiny tiny little portion of people in the *EVE* community who will take things too far. There's some crazy people out there, and it does worry me. But it's a minor worry. Not enough to keep me from [the game.]"

While there are beautiful sides of *EVE* that I've had to leave unexplored in this book, there are also ugly ones. For example, I've tried when possible to highlight women's stories within *EVE*, but the truth is that they are devastatingly few, especially in nullsec. The total player population is not even 5% women, and the reason isn't hard to figure out if you speak to enough players. Every woman I spoke to for this book told me stories of being harrassed and targeted by other players, and sometimes horrifically so. Even for the internet. The vast majority of them choose to leave *EVE*, and the community is incalculably poorer for it. The women who encounter this treatment and stay a part of the *EVE* community tend to be utterly unique people who manage to fight past harassment by sheer force of personality. One of the women mentioned in this book was a miniature donkey rancher in the real world who hunted the hilly forests of Virginia from the back of her mule. Another was a scientist who operated a nuclear reactor during the day and led fleets by night.

"I love me some *EVE* players," the miniature donkey rancher/infamous low-sec pirate lord told me. "But there are some *assholes* in this game." (Emphasis hers.)

They all told me some version of the same story: one day some jerks started harassing them, and they decided they wouldn't be pushed around anymore. They told me stories of crushing their enemies, and coming to love that feeling.

Sometimes they have to go to extreme lengths to do it, but the results can be spectacular. To my knowledge, in the history of online gaming nobody has ever gotten owned harder than the turds who doxxed Greygal, a fleet commander, newbie mentor, and community leader in high-sec. In the real world, she's a 55-year-old woman from the deep woods of the Pacific Northwest, and she wasn't about to get pushed around by some internet shitlords. So she tracked their IP numbers, found out where they lived, called their local sheriff, and as part of the plea agreement demanded that they write her a handwritten note of apology. Which they did. She still has the letters. Call it frontier justice. Not every story has justice at its conclusion though. In fact, most didn't, and it has been a powerful force in shaping the structure of the game and its community over the years.

POWER

Because *EVE* provides mechanics for exercising power over other players and other groups there exists a sort of social evolution that sculpts the community through time in all sorts of ways. The fight to collect power and stave off that evolution creates ever more complicated and capable organizations out of those who remain in the game. Organizations that provide more fun, more gain for their members, and more cultural connectedness tend to prosper and will often destroy or subdue less organized groups that they perceive threaten their goals.

That also creates a sense of urgency to avoid defeat because the punishment is the destruction of your community, and so some players will go to extreme lengths to win.

On the broad scale, the study of *EVE* is the study of a human social ecosystem in which social groups are planted, grow up, compete in their environment, and fertilize the soil for the next generation of social groups. Player groups amass a great deal of institutional knowledge through their successes and failures. When a group falls apart, its former members emerge as more confident and skilled leaders. They often take up leadership roles again in the future, and apply the lessons learned from the death of their previous organization to strengthening the new one. All toward the purpose of staving off that simple truth: Things fall apart. The center cannot hold.

When the game first launched in 2003 the largest groups numbered in the dozens and low hundreds. Today their ranks are literally legion, with numerous groups surpassing a thousand members, and the very largest numbering in the tens of thousands. Each of them are the most recent link in a chain of failed predecessors stretching back to 2003.

The big question that has hung over *EVE* for its entire existence is whether the social system of the game is an infinite conflict continuum or if it will eventually produce a winner whose control over *EVE* will be so absolute that the game breaks and the social community suffocates.

SERENITY

The only corollary to the history of the *EVE Online* Tranquility server cluster is that of its sister server, Serenity, home to the Chinese community. However, that story is a cautionary tale. I've never done a report into the events of the Serenity server, but official reports from CCP Games say it has been ruled by an overwhelmingly powerful alliance known as the Pan-Intergalactic Business Community and its ally "Veni, Vidi, Vici" (3V) since 2009. The leaders of the two coalitions are known to be close real life friends from the city of Chengdu.

There were other large alliances on the server but none that could meaningfully challenge its dominance, including "City of Angels," an alliance run by a woman who had named her character the phrase "Everyone Log Off Now" in Mandarin so that enemy fleet commanders who targeted her would have to say to their fleet "Target Everyone Log Off Now!" In 2013–roughly parallel to the events of the Halloween War but in the opposite time zone—the alliances RAC, City of Angels, and the Fadeklein Alliance tried to negotiate with the dominant PIBC for access to some backwater territory, but PIBC refused. Later, an audiolog was leaked to the community in which one of the leaders of the PIBC admitted that the purpose of the refusal was to choke off the last of the resistance to the PIBC/3V coalition.

The opposing alliances—the 2nd, 3rd, and 4th largest groups on Serenity—came together with a splinter organization of defected PIBC pilots and tried to take on the behemoth of Serenity. The assault failed,

and culminated in what became known as the Slaughterhouse at 49-U6U. On March 25, 2014, the two coalitions fought a 21-hour battle eerily reminiscent of the one at B-R5RB. Though it was interrupted by two server crashes, the PIBC/3V coalition emerged even more dominant than the ClusterFuck Coalition had. The toll of the battle eclipsed even that of B-R5RB as 84 Titans were destroyed—69 of them by PIBC/3V.

"Someone tried to defect, it didn't go well, and PIBC crushed them," said Duo Ye, who was a producer at *EVE Online*'s Chinese operator at the time, an online gaming company called Tiancity, speaking to the gaming website RockPaperShotgun.com. "Everyone is wondering now whether PIBC is so big that it's just mopping up the rest of the map and becoming the only viable alliance, but I would say not from what I see. Yes there are consequences, for example one of the losing alliances paid a heavy amount of ISK to regain friendship with PIBC."

On Serenity, a similar monument to the one that was created to honor B-R5RB instead stands in 49-U6U. In subsequent years, the PIBC's grip on power only strengthened. It's one of the reasons why in late 2013, an alliance called "Fraternity." was created on the Tranquility server. Fraternity was a home on Tranquility for Chinese players who no longer wanted to play on the Serenity server, crushed as it was by hyper-inflation of the currency and the dominance of a super coalition. During the battle at B-R5RB there were already 1500 Chinese players using VPNs to play on Tranquility as part of Fraternity. Though it was a niche alliance for several years, Fraternity's membership skyrocketed in 2018 and early 2019 reaching 12,000 players, one of the largest alliances on the modern Tranquility server. Events on the Serenity server may one day be seen as crucial forces that affected the history of Tranquility. Meanwhile, the situation on Serenity rapidly degraded. Throughout 2018 and 2019 the server was on something akin to life support often seeing fewer than 700 players online even at peak hours.

The *EVE* community is once again faced with the opportunity and the challenge of cultural fusion. In the past it's collision with the Russian community was fraught with stereotyping and name-calling. Today, with the benefit of history, it's clear that the *EVE Online* community has been greatly strengthened by the contributions of its Russian and Ukrainian players throughout the decades. It remains to be seen whether modern generations of players will be able to learn from history and avoid the mistakes of their predecessors. It's worth noting that the core of the ClusterFuck Coalition never would have achieved its station at the top of *EVE* and become The Imperium had it not been through frequent collaboration and integration with the Russian community. The future of *EVE* will belong to those who understand its past, while successfully collaborating with new and emerging cultures on Tranquility.

However, at times the reaction of established players on Tranquility has been a mirror of ignorance in the real world with a loud minority of players sneering that Chinese players on Tranquility should go back and fix their own server before they ruin Tranquility too. With cultures colliding in the real world as well, I would urge those people to consider how precious and rare an opportunity a game such as *EVE* is to engage in acts of civility en masse with members of other cultures. Never before have human beings had a communication tool—a meeting ground for common people—such as *EVE* and other modern video games and virtual worlds. Even now, history is marking how we use them.

THE MODERN ERA

The modern era of *EVE* has proved no less fascinating than the eras of *Empires of EVE: Volume 1 & 2*. Though our story ends here—for the moment—the era that follows is perhaps *EVE*'s most fascinating epoch yet, and tells the story of a secret plot to destroy The Imperium which sparked a mass community rebellion. The legacy of this famed event is still playing out, and the fight to control how it is viewed continues.

The story of Tranquility is continuing to develop even after CCP Games was purchased by South Korean gaming company Pearl Abyss in September 2018 for $425 million. Though proclamations of *EVE's* imminent demise have resounded since the earliest days, EVE ultimately shows no true sign of stopping any time soon. During the writing of this book EVE celebrated its 17th anniversary and seems poised to soon leave its teenage years behind.

The story of EVE may go on for quite some time. And if it does then we'll get new perspective into the game, identifying larger epochs and gaining a higher understanding of how digital human social systems evolve throughout time. Whatever happens, we'll learn something about what awaits humanity in its digital future.

Whether that future is a "Pan-Intergalactic Business Community" of Tranquility's own, or if the many wars described in these books will continue on forever, either would be equally fitting for a cold, dark, and harsh world.

Above: The Slaughterhouse at 49-U6U was the defining battle of the Serenity server, and may one day be seen as a factor in the history of the Tranquility server. Below: The heartbeat of *EVE Online*. The average "peak concurrent users" logged in daily from May 2003 to February 2014 courtesy of eve-offline.net.

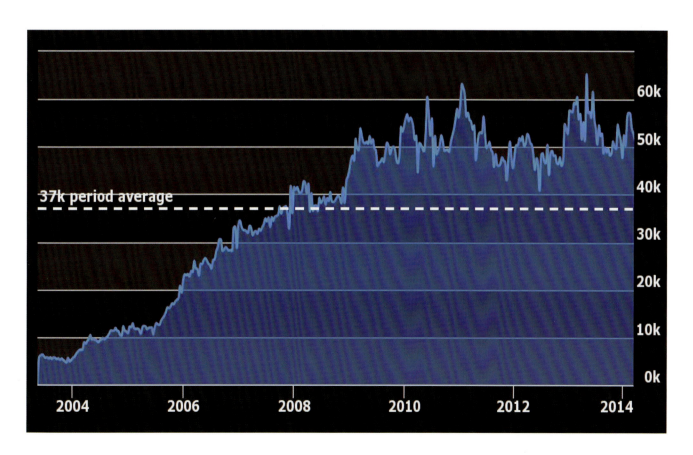